# A Democratic Approach
# To
# Sustainable Futures

## A Workbook for Addressing the Global Problematique

By
**Thomas R. Flanagan and Kenneth C. Bausch**
**Institute for 21$^{st}$ Century Agoras**

**Preface by Ambassador John W. McDonald**

*Ongoing Emergence Press*

Riverdale, GA 30274

Thomas R Flanagan and Kenneth C Bausch
A Democratic Approach to Sustainable Futures: A Workbook for Addressing the Global Problematique, collaborative project of the Institute for 21$^{st}$ Century Agoras
p.cm

Includes bibliographic references.

ISBN 978-0-9845266-1-1

   1. Group Problem Solving  2. Sustainability  3. Class Textbook  4. Global Problematique

Preface by Ambassador John W. McDonald

*Ongoing Emergence Press*

Riverdale, GA 30274

*Dedication*

*This book is dedicated to the living memory of*

*Professor Hasan Özbekhan, friend, colleague and teacher.*

*From his towering wisdom,*

*Hasan has marked us with his words "can implies ought."*

*And so armed with the wisdom of what good can yet be done in the world,*

*we are heir to the burden of doing it.*

*Acknowledgement*

*This work would not have been possible*

*without the encouragement and collaborative support*

*of Professor Janet McIntyre at Flinders University*

*and the international body of students she convenes*

*in an ongoing course on Democracy and Global Sustainability.*

*We have benefited both from inspiration and critique,*

*and carry the responsibility of an effort which, with time,*

*can still be substantially improved.*

# Preface

Theorists in International Relations try very hard to categorize and organize the behaviors of states and state leaders, and to a degree can usually reasonably demonstrate patterns which can be interpreted through the lens of various theories.  And yet surprises happen.  In fact they happen more often than not, leaving theorists to scramble for explanations consistent with theory.  The theories are not particularly useful to develop predictive models.

Systems theory grapples with the complex, acknowledges the unpredictable, and more closely reflects the interactive nature of global dynamics, including the many fields of play impacting every decision.  This workbook has the courage to take on a systems theory approach in understanding global interactions.

A Democratic Approach to Sustainable Futures: A Workbook for Addressing the Global Problematique encourages interactive learning through original research and problem exploration, and should lead students to have a sense of ownership over what they have learned after they complete the work. The authors take a systems approach that can be compared to multi-track diplomacy; dialogue, respect and understanding are all cornerstones of this piece.

The framework is built around forty-nine continuous critical problems that cannot be solved simply by isolating each problem. This encourages students to realize how complex international problems are and how difficult it is to find a good solution. Each chapter contains more homework assignments and exercises than would normally be expected, but the authors stress that information overload will help students better understand the complexity of the world in which they live as well as build community in their attempts to complete the tasks. This sets this manual apart from others which generally teach elementary concepts only.  Students are encouraged to take rational approaches to each hurdle, but the authors also ask students to rely heavily on "body wisdom", or intuition.  As any field practitioner knows rational approaches can only lead to expected outcomes.  Well honed intuition can prepare the student for more likely, less expected outcomes.  A truly skilled field practitioner is both well prepared and alert to both unexpected opportunities and dangers.

A Democratic Approach to Sustainable Futures requires students to guide their own learning. This is one of the most effective means of learning and students following this workbook can expect to feel like they have accomplished significant gains at the end.

This is an excellent workbook for serious students who are not satisfied by simple theory or impractical practice. This is the book to prepare for the messy, multi-layered, multi-faceted, personal, political, real world of applied global activism.

Ambassador John W. McDonald

Founder (1990) and CEO, the Institute for Multi-Track Diplomacy
Four-time Ambassador under Presidents Carter and Reagan
Leader of development efforts in Turkey, Iran, Pakistan, and Kenya
Past Secretary General to the Club of Rome

# Table of Contents

## Introduction

This workbook uses structured inquiry to support classroom exploration into the problematique of global sustainability. The exploration can be conducted either online in virtual classrooms or face-to-face in traditional course settings. The exploration has been designed to span four weeks of focused effort.

The course begins with a historical review of the challenges of global complexity faced by the founders of the Club of Rome circa 1970. These challenges were stated succinctly in the Club's founding prospectus. This historic prospectus introduces the *problematique,* a set of 49 continuous critical problems which interact and entangle as wicked problems. These problems are not solvable in isolation because fixing one of them will likely further complicate other problems. These problems are further delineated – although in a preliminary fashion – in Chapter 6.

The continuous critical problems are presented as topics for individual student research. As students each develop a level of expertise on the specific continuous critical problems assigned to them, the class will be confronted with a real world challenge. They will be collectively overwhelmed with information about the problems that confront global sustainability. This information overload is a critical part of the class experience. It forces a realization that we as a society are confronted with information overload. As citizens and emerging leaders in society the students need to understand approaches for dealing with this information overload in a constructive – progressive – fashion.

Students will begin by posting concise statements explaining the critical problems assigned to them along with their most preferred internet resource references. These statements express, of course, an incomplete understanding of what the student has researched. [Instructors may expand the written requirements of the course to include more complete expressions of student understandings.] The goal for the collective inquiry is to provide enough information so that the class can agree that the meanings are clear enough to proceed to further understanding

Students will collaboratively clarify the meanings of these critical problems. To do this, they will first post statements to the class wiki or the class wall to clarify the problems assigned to them. Second, they will read posted questions from other students and then post questions (as discussion threads) to those statements. Finally, each student will answer the questions raised by their classmates. This is an asynchronous group learning task. Even in face-to-face classrooms, adding reflective clarification as a "homework" assignment is important so that students have time to think about both the questions that they feel MUST be asked and the clarifications that they feel MUST be provided. In these exercises, students will discover that concise and articulate expression is important, both in class and in cross-cultural exchange as we move into the Information Age.

Students will "take ownership" of the meanings of individual continuous critical problems through their research, their descriptions, and their clarifications. "Shared ownership" of the continuous critical problems begins with clarification and then extends into tasks which specifically explore relationships among the problems. This course takes a systemic approach that avoids the trap in which group attention is focused on individual problems to the general exclusion of the relationships among problems. Systems thinking is cultivated by constructing "affinity clusters" of problems, and then by mapping "influence patterns" among problems which initially seem of greatest importance to the class. These activities include collective decisions as well as aggregated individual decisions.

The approaches presented in this course are generally applicable for engaging any sufficiently complex body of information. Engaged individuals approach a complex situation drawing on both their cognitive wisdom and their body wisdom. They are presented with opportunities to draw upon both types of wisdom in their construction of new meaning. Strategically, a community handles complex bodies of information best by using a systematic approach for understanding and reacting to the complexity. Students who take this course will come to understand this approach.

It is our hope that students will feel closer to their classmates – whether within the same academic department or when seated in classrooms across the world – and that the feeling of shared experience and shared understanding will enhance their awareness that they are members of a global community.

# Chapter 1

## Engaging Global Sustainability as the Predicament of Mankind

The 1960s were a time of upheaval and radical questioning. People questioned the way things were run and the direction the world was taking. The Club of Rome was foremost among the questioners. The directions taken by the Club of Rome influence the way we think today. Anthony Judge (March, 2010) wrote the following concerning the early days of the Club of Rome:

> The first "Report to the Club of Rome" arose from a project falling directly under the cognizance of the Executive Committee of the Club of Rome during its formative stages. The Executive Committee had asked the Institute Battelle at Geneva to provide administrative support and act as managing agency for a project Work Group and asked Hasan Ozbekhan to undertake the overall direction of the project and ensure the operational responsibility for the Work Group, calling on consultants as required to transform the prospectus into an action plan. The consultants included Alexander Christakis, Erich Jantsch, and Aurelio Peccei. The "prospectus" in the form of a "Report to the Club of Rome" was entitled: *The Predicament for Mankind: Quest for Structured Responses to Growing World-wide Complexities and Uncertainties* (1970), its proposals were rejected in favor of the "Club of Rome Project on Predicament of Mankind at MIT" directed by Dennis Meadows from 1970-1972. This resulted in publication of what became known as the first report to the Club of Rome (*The Limits to Growth,* 1972). A distinct report was later published by Hasan Ozbekhan (*The Predicament of Mankind,* in: C. West Churchman and Richard O. Mason, eds., *World Modeling: A Dialogue*, North-Holland, 1976).

This first section of this book sets the background history, the contents of Hasan's original *Predicament,* and the subsequent history of the Club of Rome and the progeny of Hasan's original efforts. Later chapters explain the methodology now available that can deal with the overwhelming complexity envisaged in Hasan's prospectus and identify organizations that are dealing with the 49 systemic problems defined in it. The software included with this book will enable students and classes to directly prioritize those problems in the fields of their interest.

**Aurelio Peccei**

Consider a time before the United States had an environmental protection agency. It was an era when the victorious military-industrial complex from the Second World War launched us on a trajectory of production for the sake of production and consumption for the sake of consumption. In the late 1960's Aurelio Peccei, an Italian industrialist, was traveling around the world trying to persuade leaders that we were facing an unprecedented global crisis. Peccei was making the case that there was an ever-growing gap between technologically developed North America and the rest of the world that posed a major global crisis between the developed and the less developed regions of the world, unless some measures were taken by world leaders to close the gap.

Peccei was involved in the resistance during World War II. He was arrested, imprisoned, tortured, and was facing execution. He escaped. After the war, he helped rebuild Fiat in Italy and Argentina. He also had strong interests in social responsibility. In 1965, he gave the keynote speech at a conference that drew the attention of significant world leaders.

Alexander King towards the end of 1966 read this speech and met Peccei in Paris. King was a distinguished research chemist and a pioneer of science in government policy in Britain during WWII. At the time of this meeting, he was Director General of Education and Science at the Organization for Economic Cooperation and Development (OECD). Peccei and King brought 30 European economists and scientists together in Rome for a two-day brainstorming meeting on April 7–8, 1968. They discussed the global nature of the problems facing mankind and of the necessity of acting at the global level. This meeting was not a success, but in an informal gathering afterwards, the group decided to call themselves the Club of Rome. They also named the cluster of intertwined problems that they would have to face as "the problematique".

> Thus started what Peccei called "the adventure of the spirit". He was fond to state that: "If the Club of Rome has any merit, it is that of having been the first to rebel against the suicidal ignorance of the human condition." Another quote of Peccei states: "It is not impossible to foster a human revolution capable of changing our present course" (Wikipedia on Peccei),

**Hasan Ozbekhan**

By 1969, Peccei was becoming increasingly impatient. The early meetings of the embryo Club had discussed problems at length but had not developed any course of action. What he was seeking was an effective methodology to tackle the issues of the problematique, which in *The Chasm Ahead* he had called "a tidal wave of global problems". In search of a structured approach to the problematique, he sought out the well-known American systems analyst Professor Hasan Ozbekhan.

Ozbekhan worked as a management consultant and was considered one of the most prominent planning theoreticians in the 70s. Peccei and Ozbekhan became very good friends, and during one of their meetings, Ozbekhan proposed to Peccei that he adopt the systems approach (Churchman, 1979) for influencing the stream of world events. This proposal was accepted in principle by Peccei in 1969. The intent was to use the findings of such a study in discussions with world leaders in order to maximize Peccei's effectiveness.

The two men agreed to develop a new strategy for the future. The Club of Rome would have members of different nationalities, different cultures, and different disciplines. In other words, the major criterion in the formation of the Club of Rome was not homogeneity, as is usually the case in most clubs, but heterogeneity. The mission of the Club of Rome would be to conceptualize the new systems approach and to fund projects by researchers around the world, all of which would focus on the impending global crisis.

An Executive Committee of the Club of Rome was formed consisting of a diverse multi-cultural group of members representing such countries as England, the Soviet Union, Germany, Austria, Switzerland, Italy, and the U.S.A. The Executive Committee commissioned Ozbekhan to write the prospectus of the Club of Rome and within six months Ozbekhan produced the first draft of the prospectus under the title "the Predicament of Mankind." At that time, Ozbekhan hired Aleco Christakis as a consultant to the Executive Committee. Other members of the research group were Erich Jantsch and Aurelio Peccei.

**Historical Account from the Club's Perspective**

Ozbekhan made his initial report to a meeting of the executive council in December, 1969. The executive meeting was held in Vienna and had two objectives. One was to give the Club of Rome legal status by incorporating it as a non-profit organization; the other was to define the problematique, with the help of Hasan Ozbekhan. The first objective was fully achieved; the second objective proved to be difficult.

> During Ozbekhan's presentation, most of those present became increasingly puzzled as to what was its purpose. It was couched in social science jargon which, while it would presumably make sense to his professional colleagues, might well cause difficulty to others.  After Ozbekhan left, Aurelio asked for comments. We told him that, learned as Ozbekhan's work no doubt was, it was not likely to be comprehensible to either the public or to politicians to whom the Club of Rome message was to be addressed. Aurelio became very upset – in fact at one point he was actually in tears – and it took several hours over dinner to pacify him by suggesting (without a lot of confidence) that he ask Ozbekhan to make another attempt at a paper on the *problematique*. Ozbekhan may have had a lot to offer through his academic work but we felt that his style did not match Peccei's vision at this point (Whitehead).

The positive outcome of the meeting was the decision to incorporate the Club of Rome as a non-profit organization under the laws of Switzerland.  The Club of Rome held its first formal Annual Meeting in Bern, Switzerland in June 1970.   The following is Whitehead's account of the meeting.

> Ozbekhan was to present his revised problematique to that meeting. It did not catch the imagination of the Members. Some were reported as saying that it was too humanistic and not structured enough. Peccei was almost at his wits' end when, towards the end of the day Professor Jay Forrester of MIT made a concrete proposal. He had previously had discussions with Peccei at MIT and he was becoming increasingly convinced that the techniques of "Industrial Dynamics" which they were successfully applying to complex industrial problems, could be adapted to model the dynamics of the world. To this end he renamed it "System Dynamics" at the suggestion of Eduard Pestel, who agreed to present the proposal to the Volkswagen Foundation for funding.   In return, Forrester invited the six Club of Rome executives to Cambridge, Mass. to discuss with him the parameters of the model.

### Historical Account from Aleco Christakis' Perspective

The Predicament of Mankind prospectus was distributed in 1970 (for a review of the prospectus visit us at the Institute for 21st Century Agoras; www.globalagoras.org). The subtitle of the document was "Quest for Structured Responses to Growing World-wide Complexities and Uncertainties." A number of prominent systems scholars of the time, such as Erich Jantsch and West Churchman acclaimed it as an outstanding prospectus. The other half, especially those trained in traditional analytical disciplines like systems engineering, thought the document was lacking in methodological rigor. Most of the latter reviewers did not realize that the prospectus was intended as an architectural design concept rather than an engineering blueprint. The distinction between an overarching architecture and a concrete plan is generally filled in by the "designer," and Hasan's approach sought to include many other voices in filling in the design. The individual expert mindset of the time was poorly prepared to appreciate the practical utility of Hasan's vision.

In retrospect, the prospectus had incorporated the seeds for a paradigm shift in designing social systems. It moved from the top-down mode of designing to a bottom up mode. Designers in the top-down mindset were unable to shift to a new paradigm. Hasan describes the need for a paradigm shift in the following passage:

> "The source of our power lies in the extraordinary technological capital we have succeeded in accumulating and in propagating, and the all-pervasive analytic or positivistic methodologies which by shaping our minds as well as our sensibilities, have enabled us to do what we have done. Yet our achievement has, in some unforeseen (perhaps unforeseeable) manner, failed to satisfy those other requirements that would have permitted us to evolve in ways which, for want of a better word, we shall henceforth call 'balanced.' It has failed to provide us with an ethos, a morality, ideals, institutions, a vision of man and of mankind and a politics which are in consonance with the way of life that has evolved as the expression of our success. Worse, it has failed to give us a global view from which we could begin to conceive the ethos, morality, ideals, institutions, and policies requisite to an interdependent world.

He goes on to relate this to the intellectual silos that isolate areas of knowledge from each other and from people living in the world.

> This failure is often regarded as having created a number of separate and discrete problems capable of being overcome by the kind of analytic solutions our intellectual tradition can so readily generate. However, the experience of the past twenty or thirty years has shown with remarkable clarity that the issues which confront us in the immediate present, as well as their undecipherable consequences over time, **may not easily yield to the methods we have employed with such success in the bending of nature to our will.**

7

Hasan goes on to enumerate some of the qualities of today's problems that might account for our ability to solve them as separate and discrete problems. He says this inability could be attributable to many things:

- the magnification of the problems we must grapple with -- that is, to the fact that almost all of them are global in scope, whereas the socio-political arrangements we have created are ill-equipped for dealing with issues that fall outside their strictly established jurisdictions.

- heightened yet often obscure interactivity among phenomena, whereas our manner of solving problems owes its strength and efficiency to the identification of rather clear and direct lines of causality.

- rapid rates of change, especially in the technological sector, whereas our institutions, outlooks and minds are geared by long-time habit to beliefs in slow unfolding and permanence -- beliefs which have sustained certain relatively stable concepts of polity, of social order and of intellectual orderliness.

"In brief," he says, "the conjuncture of events that surrounds us is to all evidence world-wide, complex, dynamic, and dangerous."

These statements, written in 1970, appear apocalyptic, and even more applicable and relevant to our contemporary situation, i.e. the beginning of the 21[th] century.

It is not our intent in this chapter to engage in revisiting the power of the concepts so masterfully described in the Club of Rome prospectus. There are, however, two aspects of the prospectus which provided the intellectual and inspirational foundations for subsequent work by many researchers in the development and validation of the Structured Dialogic Design (SDD) science paradigm (Christakis, 2006).

- The first aspect is the idea of the "value base."

- The second is the idea of the *Problematique* itself.

It is important to appreciate both of these concepts, as well as their relationship to the development of the Structured Dialogic Design as both a science and a paradigm of practice.

*The Value Base*

Hasan explained the necessity for creating a new value base in his article: Toward a General Theory of Planning. He explains the need for this new value base in the following words:

> The familiar concepts, values, thoughts, and approaches, which until very recently served to focus our perceptions, to clarify, reorder, and relate what is obscure and unmanageable in our situation are now found wanting. They are no longer able to shake down the present into some comprehensible design; that is, into a design which is meaningful in relation to current happenings. Old methods of observation, analysis, resolution and classification appear to have become obsolete. Our single-valued logic is ill-suited to probe for the multi-valued depths of the sense-data we receive; with the uncertainties that both nature and society seem to generate thorough the mere fact of existing, unfolding and evolving. The cultural referents we had come to view as absolute – liberty, equality, privacy, dignity, the individual, the nation, constitutions, the common good, and so forth – no longer provide automatic guidance to our feelings or to our behavior. Their ultimate, no less than the operational, significance is now blurred when it is not irrelevant, and insufficient when it is not confusing,
>
> \*       \*       \*
>
> We have not come within shouting distance of putting the situation aright. Whatever has lost its validity has not been replaced by any new or consistent norms. When questions are asked on this subject, the answers one gets are at best simplistic. One is told to *"Make love, not war"*; one is told of *"flower power,"* and the importance of seeing reality *"like it is"*; one is told to be *"with it"*; one is told the work, the rewards, the action, the decisions – especially the decisions – must be "shared." These haphazard injunctions do not amount to a new value system. But they tell us, with desperate insistence, that the situation has changed and that we must recognize this fact.
>
> \*       \*       \*
>
> At present, our perception is still largely governed by the world view of nineteenth century industrialism. Our problem is that the reality we are beginning to sense and with which we must deal belongs, by our own admission, to some other, newer order.

It is the rationalizing principles of this latter reality that we are now called upon to define (Ozbekhan, 1968).

Hasan recognized that this common value base requires that *dialogue has to be built into each design application.*

*The Problematique*

The *Problematique* was used in the Club of Rome proposal to draw a linguistic, and hence conceptual, distinction between the well-bounded problems that we are accustomed to perceive and articulate and the meta-problem (or *meta-system of problems*) that emerges as a result of the interactivity and interdependency among these problems. In essence, it is the equivalent concept to Churchman's conceptualization of "enormous problems."  These problems are so intertwined that solving one of them inevitably disturbs another and another, and so on.  Taken together, these problems are the world's "wicked problem" (Rittel and Weber) and the great gigantic "mess" (Ackoff).  Naming the problem as the *problematique* foreshadowed the decline of purely reductionist thinking as a dominant means of problem solving.

It is the nature of our languages, hence our manner of perceiving reality, that we see and call the dissonant elements in a situation "problems."  Although it is true that there are certain problems (mostly within one discipline and usually in the field of engineering and physical sciences) that can be addressed in their own domains by applying problem-solving techniques, it has become empirically evident over the last sixty years, that critical issues in the social arena are not capable of being solved in their own specific domains.

The Predicament prospectus identified 49 Continuous Critical Problems within the global problematique.  When we consider the truly critical issues of our times such as "environmental pollution," "poverty," "homelessness," "criminality," "population explosion," "urban deterioration," "racial and cultural discrimination," etc., we recognize that it is virtually meaningless to view these as problems that exist in isolation. The problems clustered under words such as "hunger" or "malnutrition" in Somalia cannot be separated from social, economic, and geo-political problems on the global scale. Trying to solve any of these problems in isolation exacerbates the intensity of the *Problematique*, whose solution is beyond the scope of the concepts and methods we had traditionally employed during most of the 20th century (Christakis, 2006).

The strength of couplings and overlaps among these critical problems is such that a new approach is required to model the situation as a single complex system, as opposed to the traditional means of breaking it apart in its component parts and assigning responsibility to various individuals or government departments with expertise in the separate areas. For a systems approach to be effective and meaningful, it should engage stakeholders in interdisciplinary dialogue for the purpose of articulating the *Problematique* and integrating the relevant knowledge and wisdom of the community. The approach should also redefine the notion of "the expert" so that the distinct voices of the people affected by the *Problematique* will all be heard, independent of their education, social status, or power. [Note: Every individual will not necessarily be heard, but every distinct individual perspective will be heard through some individual who will give voice to that perspective. It is the job of sponsors to make sure that all of those voices are brought to the table]

The original conceptualization of the Club of Rome prospectus advocated the position that any attempt at resolving the global *Problematique* founded on traditional elitist, exclusionary, and disciplinary approaches is doomed to failure. The majority of the Club of Rome were locked into a top-down design mindset and rejected this approach as lacking in methodological specificity and rigor.

As a result, the Executive Committee of the Club of Rome gave its support to the system dynamics group of MIT (Christakis, 2006).

**Jay Forrester and System Dynamics**

Jay Forrester was an electrical engineer at MIT. During WWII and afterwards, he designed feedback control systems for military equipment, created aircraft flight simulators, and developed MIT's first general purpose digital computer. His experiences as a manager in these situations led him to believe that social systems are harder to manage and control than are physical systems. He sought to bring science to bear on the core issues faced by corporations. His insights into the common foundations underlying engineering led to his work on industrial dynamics and his creation of system dynamics.

During the late 1950s and early 1960s, he and his students rapidly developed a system dynamics computer modeling language. Forrester published *Industrial Dynamics* in 1961. In 1968, he branched beyond corporate applications into urban planning and wrote *Urban Dynamics*. In response to the notoriety of that book, Hasan Ozbekhan and Aleco Christakis invited him to a plenary meeting of the Club of Rome in Bern Switzerland.

At this meeting after Hasan's proposal met with a largely unsympathetic reception, the Club turned to Jay Forrester and asked him if he could adapt the principles of urban dynamics to the global problematique.  He assured them that he could.  Three weeks later, a group of Club members visited Forrester at MIT and were convinced that the model could be made to work for the kind of global problems which interested the Club. The Executive Committee sponsored the production of a "world model" using the methodology of system dynamics.  The work and findings of this project culminated with the publication of the very popular book *Limits to Growth* in 1972 (Meadows, et al., 1972).  The controversial nature of the findings reported in this book gave a lot of publicity and notoriety to the Club of Rome.

The history of the Club of Rome demonstrates the power and weakness of its Systems Dynamic approach to global problems.  Such research has developed strong models of global systems "using a highly restricted number of key variables" (Judge).  The Club's research has "focused on influencing decision makers and power brokers through meetings" (Judge).  The weakness of the system dynamics approach was demonstrated by the United Nations Climate Change Conference (Copenhagen, 2009). In the words of Anthony Judge:

> It is not what some assume to be the "facts" that elicits a political consensus or reflects the actual dynamics of a global psychosocial system.  The challenge remains one of engaging with divergent "perceptions" and determining what scope there is for configuring them more fruitfully.

It is precisely the need to deal with divergent perspectives that is the strength of Hasan's *Predicament* and the methodology of Structured Dialogic Design® (SDD).  SDD takes the time to build respect and understanding among participants.  Then it builds consensus among the participants about the nature and root causes of a particular situation.  Only then do the participants list their action options and finally decide upon a course of action.

**Comparison of the Methodologies proposed to the Club of Rome in 1970**

|  | **Limits to Growth** | **Prospectus** |
|---|---|---|
| Variables Considered | 5 | 49 |
| Participants | Experts | Representative Stakeholders |
| Method | System Dynamics | Successive committees of stakeholders |
| Club of Rome Decision | Accepted | Rejected |
| Positive Reaction since 1972 | Club of Rome consistently employed System Dynamics as its method of choice<br>Worldwide debate on topics such as climate change | Method envisaged in prospectus refined, resulting in Structured Dialogic Design, an effective method for dealing with complex psychosocial problems such as above 49 |
| Negative Reaction since 1972 | Rejection of results:<br>by scientists on the basis f insufficient variety of studies<br> by public for a variety of reasons | Structured Dialogic Design not widely recognized and with limited influence |
| Prognosis | Continuing dissent and change occurring usually because of catastrophic events | Gradual adoption of structured dialogue and eventual consensus on controversial issues |

When the Executive Committee made this grant award to MIT to develop the world model, Hasan and Aleco resigned from the Club. They both felt that the system dynamics methodology, which was used to derive an extrapolated future for the world system to the year 2150, compromised the original intent of the Club of Rome proposal which was to discover and use *a methodology capable of engaging the stakeholders in a dialogical process with sensitivity to their cultural situation and the praxis of their lives.* They felt that the system

13

dynamics approach was perpetuating a paradigm of scientific elitism and social engineering in designing social systems, instead of legitimizing the wisdom of the people by engaging stakeholder in a dialogue for designing their futures. Hasan joined the Wharton School of the University of Pennsylvania, and Aleco got involved with the establishment of the Academy for Contemporary Problems with the financial support of the Battelle Memorial Institute.

Aleco working with John Warfield and many other colleagues, initially at Battelle, and later in other academic institutions and in private practice, has spent approximately 35 years of research, development, and testing in the arena to create the model and methodology that renders the original architecture of the Club of Rome proposal usable and applicable in the field of practice.

The fruit of that research and practice is the Structured Dialogic Design process, which is reported in *How People Harness their Collective Wisdom and Power to Construct the Future in Colaboratories of Democracy*. *www.harnessingcollectivewisdom.com*.

---

STUDY QUESTIONS:

Do you feel that the tools and approaches we use to make decisions have a profound impact on the way that we shape the world today? Explain your answer.

When something needs to be done, is it always better to do doing something than to do nothing? Explain your answer.

Forecasting is a tricky business because it uses the behavior of the past and "mindlessly" yet mathematically projects it into the future? How can forecasters get clues that past trends are likely to change in the future?

People who are presented with a complex plan may endorse it, accept it, or challenge it. What do you feel the social consequences are for individuals who raise their voices to challenge complex plans that are presented to them? Can you make references from your own life experiences?

The story of the launch of the Club of Rome is not a story of heroes and villains. It is a story of competing decision making approaches. If the Club of Rome were being launched today, do you think that it may have taken a different course? Explain your reasoning.

---

HOMEWORK ASSIGNMENT

Each individual in the class is asked to become an "expert" for a subset of the Continuous Critical Problems (Chapter 6). Your instructor will assign specific problems to you. The random nature of the assignments contributes to the amount of learning that each participant will undertake. This will challenge some students. Some will say "If we all become expert in different things, then how can we possibly be graded in a class like this?" The answer lies in the quality of each student's explanations and clarifications of their assigned problems. Secondarily, student performance will be measured by the questions they raise and the relationships they help the group identify. And as we will see shortly, performance can be assessed on the quality of the narratives students construct to explain the group understanding of the problematique. To maximize student participation without overburdening them, individual workloads for independent study should be no more than five and no less than three problems.

---

EXERCISE: **Take "ownership" of your set of assigned problems and begin to research their meaning.**

When you have been assigned your set of problems, turn first to Chapter 6 for some hints of what your problems may "mean." Meaning is inferred by your understanding of the label for the problem. You should not feel constrained to use or reuse specific examples in Chapter 6. It is reasonable for us to presume that you will have contemporary examples to draw upon as you when you engage this task.

## Chapter 2

## Structural Inquiry Applied to the Predicament of Mankind

THE OPPORTUNITY AND NEED FOR A SYSTEMS UNDERSTANDING

In the last decade of the old millennium, Aleco Christakis and Hasan Ozbekhan sat down one April afternoon and reflected on what had and had not happened since they became involved in addressing global challenges. They wondered whether new methods for dealing with very complex problems like the problematique would have made a difference in 1970.

When the Club of Rome was launched in the early 70s, tools for managing interdisciplinary discovery work on enormous problems were just emerging. The Predicament of Mankind presented an unwieldy list of 49 distinct issues that together constitute the "problematique" of global sustainability. Largely because of this lack of tools, the Club chose the method of System Dynamics, which offered promise of dealing with at least some of those problems.

In 1995, the methodology of Interactive Management (IM) had been developed and validated. Hasan and Aleco decided to approach the original 49 problems using these new procedures. This chapter describes what they did. In so doing, it also illustrates a way of looking at a complex set of issues that are part of a single huge and messy problem. Using IM, Hasan and Aleco were able to develop a focus for concentrated action.

A COMPLEX UNDERSTANDING NEEDS TO UNFOLD

Many kinds of factors enter into a complex problem. A rich understanding of each of them will unfold only through group dialogue. Preliminary understanding comes when an idea is introduced and its name is added to a list. Additional understanding comes as an idea is clarified.

For the purposes in this text we begin with Hasan's list. In doing this, we begin downstream at a point where Hasan has done the leg work of extracting a list of 49 continuous critical problems that constitute the essence of the global problematique. The Predicament of Mankind did not include an appendix clarifying the meaning of the 49 continuous critical problems for its original audience; however, you will find clarification of meanings behind the names of continuous critical problems in Chapter Six of this book.

A list of the 49 Continuous Critical Problems ("CCPs" or "problems") identified by Hasan is presented in Table 1.  Clarifying and managing such a list is a phase of dealing with a complex problem that consumes a great deal of time.  Not only might the material itself be obscure, but the language needed to penetrate that obscurity may not yet exist within the community dealing with the problem.  At the same time, if we give in to the reductionist strategy of breaking a problem down into its component parts in hopes of understanding and fixing that problem, we generally will fail to come up with an approach appropriate for the problem in its entirety.  It is for this reason that this sustainability workbook includes an entire section of the book devoted to "approximating" a contemporary meaning for each of the 49 "problems" recognized by Hasan.

Perhaps there are more yet unmentioned Continuous Critical Problems?  However, an exhaustive review of Hasan's list has convinced the Institute for 21st Century Agoras that Hasan has indeed captured an enduring list of significant problems which even in the most contemporary setting may still prove to be an exhaustive list.  If after you have studied Hasan's list as presented in this book you feel you have discovered a new Continuous Critical Problem that should be included, write to us and if we agree we will recognize and include your contribution in the next edition of this text.

For the purposes of understanding a systematic approach to dealing with the complexity of sustainable situations on a global level, let us agree for the moment that Hasan has provided us with the essentials.  This list is shown in Table 1 and then more fully explored in Chapter 6.

---

**Table 1.  Continuous Critical Problems (CCPs): An Illustrative List**

- (CCP- 1) Explosive population growth with consequent escalation of social, economic, and other problems.
- (CCP- 2) Widespread poverty throughout the world.
- (CCP- 3) Increase in the production, destructive capacity, and accessibility of all weapons of war.
- (CCP- 4) Uncontrolled urban spread.
- (CCP- 5) Generalized and growing malnutrition.
- (CCP- 6) Persistence of widespread illiteracy.
- (CCP- 7) Expanding mechanization and bureaucratization of almost all human activity.
- (CCP- 8) Growing inequalities in the distribution of wealth throughout the world.
- (CCP- 9) Insufficient and irrationally organized medical care.
- (CCP-10) Hardening discrimination against minorities.
- (CCP-11) Hardening prejudices against differing cultures.

- (CCP-12) Affluence and its unknown consequences.
- (CCP-13) Anachronistic and irrelevant education.
- (CCP-14) Generalized environmental deterioration.
- (CCP-15) Generalized lack of agreed-on alternatives to present trends.
- (CCP-16) Widespread failure to stimulate man's creative capacity to confront the future.
- (CCP-17) Continuing deterioration of inner-cities or slums.
- (CCP-18) Growing irrelevance of traditional values and continuing failure to evolve new value systems.
- (CCP-19) Inadequate shelter and transportation.
- (CCP-20) Obsolete and discriminatory income distribution system(s).
- (CCP-21) Accelerating wastage and exhaustion of natural resources.
- (CCP-22) Growing environmental pollution.
- (CCP-23) Generalized alienation of youth.
- (CCP-24) Major disturbances of the world's physical ecology.
- (CCP-25) Generally inadequate and obsolete institutional arrangements.
- (CCP-26) Limited understanding of what is "feasible" in the way of corrective measures.
- (CCP-27) Unbalanced population distribution.
- (CCP-28) Ideological fragmentation and semantic barriers to communication between individuals, groups, and nations.
- (CCP-29) Increasing a-social and anti-social behavior and consequent rise in criminality.
- (CCP-30) Inadequate and obsolete law enforcement and correctional practices.
- (CCP-31) Widespread unemployment and generalized under-employment.
- (CCP-32) Spreading "discontent" throughout most classes of society.
- (CCP-33) Polarization of military power and psychological impacts of the policy of deterrence.
- (CCP-34) Fast obsolescing political structures and processes.
- (CCP-35) Irrational agricultural practices.
- (CCP-36) Irresponsible use of pesticides, chemical additives, insufficiently tested drugs, fertilizers, etc.
- (CCP-37) Growing use of distorted information to influence and manipulate people.
- (CCP-38) Fragmented international monetary system.
- (CCP-39) Growing technological gaps and lags between developed and developing areas.
- (CCP-40) New modes of localized warfare.
- (CCP-41) Inadequate participation of people at large in public decisions.
- (CCP-42) Unimaginative conceptions of world-order and of the rule of law.
- (CCP-43) Irrational distribution of industry supported by policies that will strengthen the current patterns.
- (CCP-44) Growing tendency to be satisfied with technological solutions for every kind of problem.
- (CCP-45) Obsolete system of world trade.
- (CCP-46) Ill-conceived use of international agencies for national or sectoral ends.

- (CCP-47) Insufficient authority of international agencies.
- (CCP-48) Irrational practices in resource investment.
- (CCP-49) Insufficient understanding of Continuous Critical Problems, of their nature, their interactions and of the future consequences both they and current solutions to them are generating.

## COMPLEX MEANINGS EMERGE THROUGH COMPARISONS AND DISTINCTIONS

A third level of understanding emerges when we compare common features of ideas. As we make comparisons of similarity, we can organize or cluster those ideas (or "problems") into "affinity groups." This clustering process starts out slowly, but as the clusters begin to grow, the distinctions become easier to make and the similarities become easier to grasp.

## AVOIDING THE TRAP OF PRE-CONSTRUCTED UNDERSTANDINGS

This is a process of putting ideas together into bins, but an important distinction is to avoid using labels for the bins before the process starts. Using labeled bins causes groups to constrain their thinking into previous ways of thinking rather than to actually discover new features of a set of ideas (or "problems") directly.

It should be noted that all three of these first steps in the exploration of complex meaning (generating a list of ideas that we feel are important, expressing what we feel to be the essence of each idea, and clustering ideas into affinity groups) we rely upon our impressions, feelings, and intuitions – in short, we rely upon our "body wisdom". As we do this, we respect each other's observations as a sincere expression of another individual's "body wisdom." We respect the words that they use as they seek to share meaning with us, without trying to adapt their words to fit some preconceived and potentially compromised understanding of their expressed meaning. It is for this reason that we start to organize ideas into groups without preordained names or labels for categories. Instead, in a systemic manner we simply ask each other if we intuitively relate one idea to another. Hasan and Aleco followed this process in 1993. They clustered problems based on similarities and discovered a set of 10 clusters which they then named.

**Table 2. Clusters of Similar Problems**

1. Population Growth / Distribution
2. Poverty, Lags and Gaps

---

3. Warfare
4. Urbanization
5. Education
6. Institutional Arrangements
7. Prejudices
8. Unknowns
9. Environment
10. Value-Base

---

## AVOIDING THE TRAP OF ONLY COUNTING WHAT CAN BE COUNTED

In what is called a "meta analysis" of the clusters – that is an analysis of the analysis itself – we can recognize two generic types of clusters appear to have been identified: Clusters 1, 2, 3, & 9 contain 24 "problems" that can be measured to some extent with quantitative tools; clusters 4, 5, 6, 7, 8 and 10 are charged with social values and cannot be objectively quantified. Under some circumstances, decision makers can be drawn toward the types of problems that can be quantified most readily – forgoing the challenges of working with subjectively measured problems. The clustering process reveals different aspects – or dimensions – of a complex situation, and in doing this clustering serves to remind planners that a balanced, multi-dimensional understanding of a complex situation will need to address multiple dimensions of the situation.

## CLUSTERING EXPLICITLY EXPOSES US TO SHARING BODY WISDOM

Hasan and Aleco placed nine "problems" into Cluster #1: _Population Growth / Distribution_ (see Table 3). As we look at the "problems in Cluster #1, we do not have a record of the deliberations that occurred and therefore can only presume how Hasan and Aleco concluded that those problems belonged in the same cluster. Our own deliberations may lead us to different conclusions. We might infer that Hasan and Aleco felt there is a gestalt shared by the problems with respect to equitable distribution of limited resources across expanding populations. The cluster of problems seems to be related to social ills that are generated as a result of official policies in direct ways.

Consider now the four "problems" that Hasan and Aleco placed in Cluster #2: _Poverty Lags and Gaps_. Again, we don't have a transcript of the reasons – or the warrants – that were applied by Hasan and Aleco to make the cluster assignments. We might see these problems as social ills that are secondary or tertiary results of official policy decisions. If we apply our own individual "body wisdom" to this task, we might feel that some of the problems would fit better if they

21

were re-clustered. We might discuss the specific features of the problems to explain why we felt that the fit were improved. We might – at some point – get the "gut feeling" that the fit was *close enough*.

Why does it matter? Why is a "gut feeling" so very important when a group is discussing a complex issue? Why is it important to legitimize our sense of body wisdom so that we can access it and share it?

The answer is that complex social issues involve making judgments and comparisons when the simplistic business of counting doesn't work. We must respect body wisdom, accept the legitimacy of body wisdom in others, and then also respectfully challenge the body wisdom of each other so that we can learn from each other things that numbers alone cannot teach. For example: Let us ask rhetorically (because Hasan and Aleco cannot inject their thinking into our question at this moment), would "problem" *Inadequate Shelter and Transportation* fit better in Cluster #2 than if fits in Cluster #1. This is a complex decision making task, and your "gut feeling" is going to help you even before your conscious thinking can put words to your choice. Actually, you will find yourself (feel yourself) making the decision first and then reaching for words to express the decision that you have made. This is natural. We are presenting an ad hoc rationalization for our feelings. When we share these ad hoc rationalizations, we may rethink our gut feelings. What happens is that the exchange of expressions may evoke some deeper wisdom within us, and this may soften or strengthen our feelings with respect to our decisions.

---

**EXERCISE**

Time yourself as you make your decision to the question presented in the prior paragraph. Where will you place *Inadequate Shelter and Transportation*? You are making an individual assessment. How strongly do you feel about your decision? Score this on a basis of 1 to 10. Next, make this same decision with a group of four of your friends, peers, and/or colleagues. Efficiency requires that you all have a shared understanding of the meaning of the problems that are in the two clusters. Go ahead and make the group determination. Time your decision making process. Score the strength of your decision. What does this exercise tell you about sharing "body wisdom" for dealing with a complex decision? Do you believe that shared wisdom is always wisdom that is stronger? What does this suggest to you about groups who need to share understandings so that they will have sufficient confidence to also share their resources in a collaborative effort?

---

**Table 3. Representative Grouping of 49 Problems into Affinity Clusters**
### CLUSTER #1: POPULATION GROWTH / DISTRIBUTION (9 ideas)
- (CCP-1) Explosive Population Growth with Consequent Escalation of Social, Economic, and Other Problems
- (CCP-8) Growing Inequalities in the Distribution of Wealth throughout the World
- (CCP-19) Inadequate Shelter and Transportation
- (CCP-20) Obsolete and Discriminatory Income Distribution System(S)
- (CCP-27) Unbalanced Population Distribution
- (CCP-31) Widespread Unemployment and Generalized Under-Employment
- (CCP-32) Spreading "Discontent" throughout Most Classes of Society
- (CCP-43) Irrational Distribution of Industry Supported by Policies that Will Strengthen the Current Patterns
- (CCP-48) Irrational Practices in Resource Investment
### CLUSTER #2: POVERTY, LAGS & GAPS (4 ideas)
- (CCP-2) Widespread Poverty throughout the World
- (CCP-5) Generalized and Growing Malnutrition
- (CCP-9) Insufficient and Irrationally Organized Medical Care
- (CCP-39) Growing Technological Gaps and Lags between Developed and Developing Areas
### CLUSTER #3: WARFARE (5 ideas)
- (CCP-3) Increase in the Production, Destructive Capacity, and Accessibility of All Weapons of War
- (CCP-29) Increasing A-Social and Anti-Social Behavior and Consequent Rise in Criminality
- (CCP-30) Inadequate and Obsolete Law Enforcement and Correctional Practices
- (CCP-33) Polarization of Military Power and Psychological Impacts of the Policy of Deterrence
- (CCP-40) New Modes of Localized Warfare
### CLUSTER #4: URBANIZATION (2 ideas)
- (CCP-4) Uncontrolled Urban Spread
- (CCP-17) Continuing Deterioration of Inner-Cities or Slums
### CLUSTER #5: EDUCATION (3 ideas)
- (CCP-6) Persistence of Widespread Illiteracy
- (CCP-13) Anachronistic and Irrelevant Education
- (CCP-37) Growing of Distorted Information to Influence and Manipulate People
### CLUSTER #6: INSTITUTIONAL ARRANGEMENTS (9 ideas)
- (CCP-7) Expanding Mechanization and Bureaucratization of Almost All Human Activity
- (CCP-25) Generally Inadequate and Obsolete Institutional Arrangements
- (CCP-34) Fast Obsolescing Political Structures and Processes
- (CCP-38) Fragmented International Monetary System
- (CCP-41) Inadequate Participation of People Large in Public Decisions
- (CCP-42) Unimaginative Conceptions of World-Order and of the Rule of Law
- (CCP-45) Obsolete System of World Trade
- (CCP-46) Ill-Conceived Use of International Agencies for National or Sectoral Ends

- (CCP-47) Insufficient Authority of International Agencies
  **CLUSTER #7: PREJUDICES (3 ideas)**
- (CCP-10) Hardening Discrimination against Minorities
- (CCP-11) Hardening Prejudices Against Differing Cultures
- (CCP-28) Ideological Fragmentation and Semantic Barriers To Communication Between Individuals, Groups, And Nations
  **CLUSTER #8: UNKNOWNS (2 ideas)**
- (CCP-12) Affluence and Its Unknown Consequences
- (CCP-49) Insufficient Understanding of Continuous Critical Problems, their Nature, their Interactions and the Future Consequences that their Current Solutions are Generating
  **CLUSTER #9: ENVIRONMENT (6 ideas)**
- (CCP-14) Generalized Environmental Deterioration
- (CCP-21) Accelerating Wastage and Exhaustion Of Natural Resources
- (CCP-22) Growing Environmental Pollution
- (CCP-24) Major Disturbance of the Globe's/World's Physical Ecology
- (CCP-35) Irrational Agriculture Practices
- (CCP-36) Irresponsible Use of Pesticides, Chemical Additives, Insufficiently Tested Drugs, Fertilizers, Etc.
  **CLUSTER #10: VALUE-BASE (6 ideas)**
- (CCP-15) Generalized Lack of Agreed-On Alternatives to Present Trends
- (CCP-16) Widespread Failure to Stimulate Man's Creative Capacity to Confront the Future
- (CCP-18) Growing Irrelevance of Traditional Values and Continuing Failure to Evolve New Value Systems
- (CCP-23) Generalized Alienation of Youth
- (CCP-26) Limited Understanding of What Is "Feasible" in the Way of Corrective Measures
- (CCP-44) Growing Tendency to be satisfied with Technological Solutions for Every Kind of Problem

## UNDERSTANDING BODY WISDOM WITH REFERENCE TO SYSTEMS SCIENCE

System Dynamics and other mathematical methods have little success in dealing with the psychological realities such as prejudice and different value bases. They are unreliable where technical systems and social systems jointly shape impacts that affect individuals, populations and ecosystems. They fail when dealing with the effects of urbanization, management styles, and unknown variables. To deal with these variables reliance upon objective observer-independent data is a formula for failure. What is needed is a methodology that deals with subjective and "observer-dependent" observations.

Interactive Management (IM) and its refinement, Structured Dialogic Design (SDD), rely upon subjective convictions, intuitions, and feelings. By bringing into the open our felt observations, intuitive clarifications, and judgments about the similarities among observations, SDD participants tap their body wisdom. They do not abandon logic, but they hold it in abeyance in order to express the subjective depth of their understanding. The subsequent stages of SDD rely upon subjective intuition in much the same way.

VOTING FOR WHAT INITIALLY SEEMS TO BE HIGHLY IMPORTANT

Having clustered the "problem," Hasan and Aleco proceeded to vote for the "problems" that they felt were of greatest importance – once again relying upon body wisdom. As a result, they selected 24 "problems" as being more salient than the others – which is to say just less than half of Hasan's exhaustive list of problems were judged to be salient. The word salient is particularly apt because it means the impression that jumps out at you or that stands out from among other options. In large groups, the process of identifying salience is facilitated by limiting selections of salience to five ideas per individual and tallying up the individual votes to an aggregate score. A subtle point here is that this type of voting is not a group decision but rather a collection of individual decisions – whereas the clustering was a decision that was generated through the group acting and interacting in collaboration.

The 24 problems that Hasan and Aleco jointly voted as being most salient are listed in Table 4. In their selection of these highly salient "problems," Hasan and Aleco used their intuition, their understanding of the problems, and their impression of the state of the world at the time that they constructed their view of the structure. Thus, their combined selection is a deeply informed selection that ultimately represents their unique "observer-dependent" understanding. As we repeat this task now standing in the future with new decision makers, some of the problems that were judged to be most salient twenty years ago are almost certain to have given way to others, and some problems that were recognized as authentic yet not seen as salient earlier may now have leapt to the forefront. The point is that a contemporary group's selection of salient problems is expected to differ both from the choices made by Hasan and Aleco and also from any parallel group operating with a distinct pool of body wisdom.

MARRYING BODY WISDOM WITH SYSTEM THINKING

Hasan and Aleco did not stop their investigation of the global problematique of the predicament of mankind at this stage because they recognized that the importance of problems judged by individual body wisdom alone is incomplete. Intuitive importance does not necessarily identify highly influential problems. Influential problems are recognizable only when those same problems are first understood in the context of the influence of problems upon each other. Hasan and Aleco were seeking a shared systems understanding of the set of problems.

To illustrate this point, consider how an important problem such as dealing with _antisocial behavior and criminal activity_ might not be very helpful in dealing with _anachronistic and irrelevant education_ -- or with the growing _tendency to be satisfied with technological solutions for every kind of problem_. Conversely, resolving the generalized _lack of agreed-on alternatives to present trends_ may not seem to be of the utmost of importance, but progress in dealing with it would definitely help us deal with a number of other problems such as _uncontrolled urban spread_ and _criminal activity_.

The tendency to concentrate efforts on "important problems" instead of "influential problems" results in wasted effort through **erroneous priorities**.

---

**Table 4. Saliency Established for 24 Problems by Preference Voting**

      **CLUSTER #1: POPULATION GROWTH / DISTRIBUTION (3 of 9 ideas)**
• (CCP-1) Explosive Population Growth with Consequent Escalation of Social, Economic, and Other Problems
• (CCP-8) Growing Inequalities in the Distribution of Wealth throughout the World
• (CCP-20) Obsolete and Discriminatory Income Distribution System(S)

      **CLUSTER #2: POVERTY, LAGS & GAPS (1 of 4 ideas)**
• (CCP-9) Insufficient and Irrationally Organized Medical Care

      **CLUSTER #3: WARFARE (3 of 5 ideas)**
• (CCP-3) Increase in the Production, Destructive Capacity, and Accessibility of All Weapons of War
• (CCP-29) Increasing A-Social and Anti-Social Behavior and Consequent Rise in Criminality
• (CCP-40) New Modes of Localized Warfare

      **CLUSTER #4: URBANIZATION (1 of 2 ideas)**
• (CCP-4) Uncontrolled Urban Spread

      **CLUSTER #5: EDUCATION (1 of 3 ideas)**
• (CCP-13) Anachronistic and Irrelevant Education

      **CLUSTER #6: INSTITUTIONAL ARRANGEMENTS (5 of 9 ideas)**

- (CCP-7) Expanding Mechanization and Bureaucratization of Almost All Human Activity
- (CCP-25) Generally Inadequate and Obsolete Institutional Arrangements
- (CCP-41) Inadequate Participation of People Large in Public Decisions
  - (CCP-42) Unimaginative Conceptions of World-Order and of the Rule of Law
- (CCP-46) Ill-Conceived Use of International Agencies for National or Sectoral Ends


### CLUSTER #7: PREJUDICES (2 of 3 ideas)
- (CCP-11) Hardening Prejudices Against Differing Cultures
- (CCP-28) Ideological Fragmentation and Semantic Barriers To Communication Between Individuals, Groups, and Nations

### CLUSTER #8: UNKNOWNS (1 of 2 ideas)
- (CCP-49) Insufficient Understanding of Continuous Critical Problems, their Nature, their Interactions and the Future Consequences that their Current Solutions are Generating

### CLUSTER #9: ENVIRONMENT (3 of 6 ideas)
- (CCP-14) Generalized Environmental Deterioration
- (CCP-21) Accelerating Wastage and Exhaustion of Natural Resources
- (CCP-35) Irrational Agriculture Practices

### CLUSTER #10: VALUE-BASE (4 of 6 ideas)
- (CCP-15) Generalized Lack of Agreed-On Alternatives to Present Trends
- (CCP-18) Growing Irrelevance of Traditional Values and Continuing Failure to Evolve New Value Systems
- (CCP-23) Generalized Alienation of Youth
- (CCP-44) Growing Tendency to be Satisfied with Technological Solutions for Every Kind of Problem

STRUCTURING RELATIONS AMONG PROBLEMS INTO A SHARED UNDERSTANDING

To determine the relative influences in these 24 problems, Hasan and Aleco used an approach called Interpretive Structural Modeling (ISM) which is central to the Interactive Management and the Structured Dialogic Design methodologies. Hasan and Aleco systematically applied the following question to each of the salient problems in a pair wise fashion: *"Supposing that humankind was able in the past two decades to make progress in the resolution of problem #2, would this progress SIGNIFICANTLY improve the capacity to make progress in resolving problem #1, in the context of the global Problematique?"*

Software provided with this book (or available through a download from the Institute for 21$^{st}$ Century Agoras website) will track the answers to this question, and will additionally infer the next pair wise comparison that needs to be made. This will save ~70% of the time required to complete the ISM task.

If the answer to the structuring question shown above is "yes," then problem #2 is placed below problem #1. The vertical position of the problems with respect to each signifies which problem is deeper down within a "tree of influence" that will emerge from these comparisons.

If the answer to the structuring question shown above is "no," then problem #2 would not yet be mapped in the structure. Its position would remain unknown until its influence upon some other problem was identified.

---

**EXERCISE**

In a group of twenty students, ask the following question: *"Will understanding **how a pair of systems scientists used some specialized approaches to understand a complex problem** SIGNIFICANTLY help us decide **how to approach really complex situations** as we look for sustainable ways to live together on the planet?"* Keep track of the time. First ask for a show of hands. Count the YES votes. Then count the NO votes. If you have folks who do not know how to vote on this question, tell the group that you need to wait for a few more moments, and then you will ask for votes again.

Body wisdom is sensed in its own time. When you have your votes, then ask for help understanding "why." Ask this question first from someone in the group of the YES/NO votes that was smaller. If you get fewer NOs, then first ask for help understanding why the answer might be NO for the group. Repeat with a response from someone from the larger of the two vote groups. Ask if anyone else would like to offer a distinct reason for their vote. Continue until the "strong opinions" have been expressed. Now vote again and count once more. If 66% of the class (e.g., thirteen or more in a group of twenty) agrees that they see a SIGNIFICANT influence, record the group response as a YES. Otherwise record the group response as a NO. "No" only means that the group did not see the SIGNIFICANT RELATIONSHIP and not that a significant relationship does not exist beyond the collective wisdom of the group. Now discuss your experience:

How long did this decision take?

Did the voting pattern change as a result of the discussion?

---

Did votes change in both directions ... YESs going to NO and NOs going to YES votes?

Did the possibility of some relationship seem obvious to all?

Did the discussion challenge the feeling about the "SIGNIFICANCE" of the relationship?

Do you feel that a group could learn to use this discussion approach better with practice?

Do you feel that this approach might be useful with a large group of ideas?

Do you feel that the pair wise comparisons, the votes, and the reasons all need to be tracked and reported?

What does the group feel would be necessary to convince the group to apply the approach to other complex situations?

Using the generic structuring question shown in the prior paragraphs, Hasan and Aleco deliberated as to whether "addressing problem #1 significantly helps in addressing problem #2." Once again, we do not have a record of these deliberations to share with you. We do not have a record of what Hasan and Aleco came to understand as being "significant" in their experiences. We do not know the extent to which perspectives differed. What we can share with you is a record of the view of the system of salient issues that existed at the heart of the predicament of humanity problematique in the understanding that was shared by Hasan and Aleco at that time.

After some three hours of deliberation Hasan and Aleco answered 120 pair wise questions related to 24 salient problems and produced the problem "tree" (or more formally the Interpretive Structural Model) shown graphically in Figure 1. The tree provides a clear picture of the way that Hasan and Aleco viewed the salient structure of the Problematique of the predicament of humanity. By convention, the tree is "read" from the bottom up. This is to say, that the story about the structure generally begins with identification of what the tree's designers have collectively agreed to be the "root cause" of the system of salient problems that define the situation.

Will the tree that your group constructs differ from the tree that Hasan and Aleco constructed twenty years ago? Almost certainly, yes. Will you find the same "root cause?" Maybe not. Some groups, for example, have come to understand the emergence of the sustainability movement as a reflection of a change in traditional value systems or as an introduction of a new value system. Your tree may be quite different. What your tree will allow you each to do is to construct an individual narrative or story about what you – as a group – have come to see as the structure of the problem that we face as a species at this moment in time. The interesting feature to consider when comparing trees is the extent to which there is agreement, not the extent to which there is disagreement. The tree, itself, (Figure 1) is an artifact showing where strong agreement exists within a group. As we discover where strong agreement exists among sets of trees, we also discover where strong agreement exists among groups. This larger knowledge is powerful when we ponder where and how large groups might begin to work together to solve global problems.

---

**EXERCISE**

In preparation for your own group's work, take turns using the "tree" representing Hasan and Aleco's shared understanding to present your own view of that understanding. You don't need to tell the entire story. Practice telling portions of the story accurately. Do you feel that the combined use of the tree and the story enhances your understanding of Hasan and Aleco's thinking? Do you feel that a visual image is an important part of understanding complex situations? Is the tree that Hasan and Aleco constructed "too complex" to be useful to you?

Recognize that accurately telling any groups story is a very difficult challenge and that only recently have tools emerged that now allow groups to capture their story so that it can be accurately told by others. Ask yourself if improved capacity in understanding and telling a group's story might have the essential capacity to guide us to a sustainable world.

# A Democratic Approach to Sustainable Futures

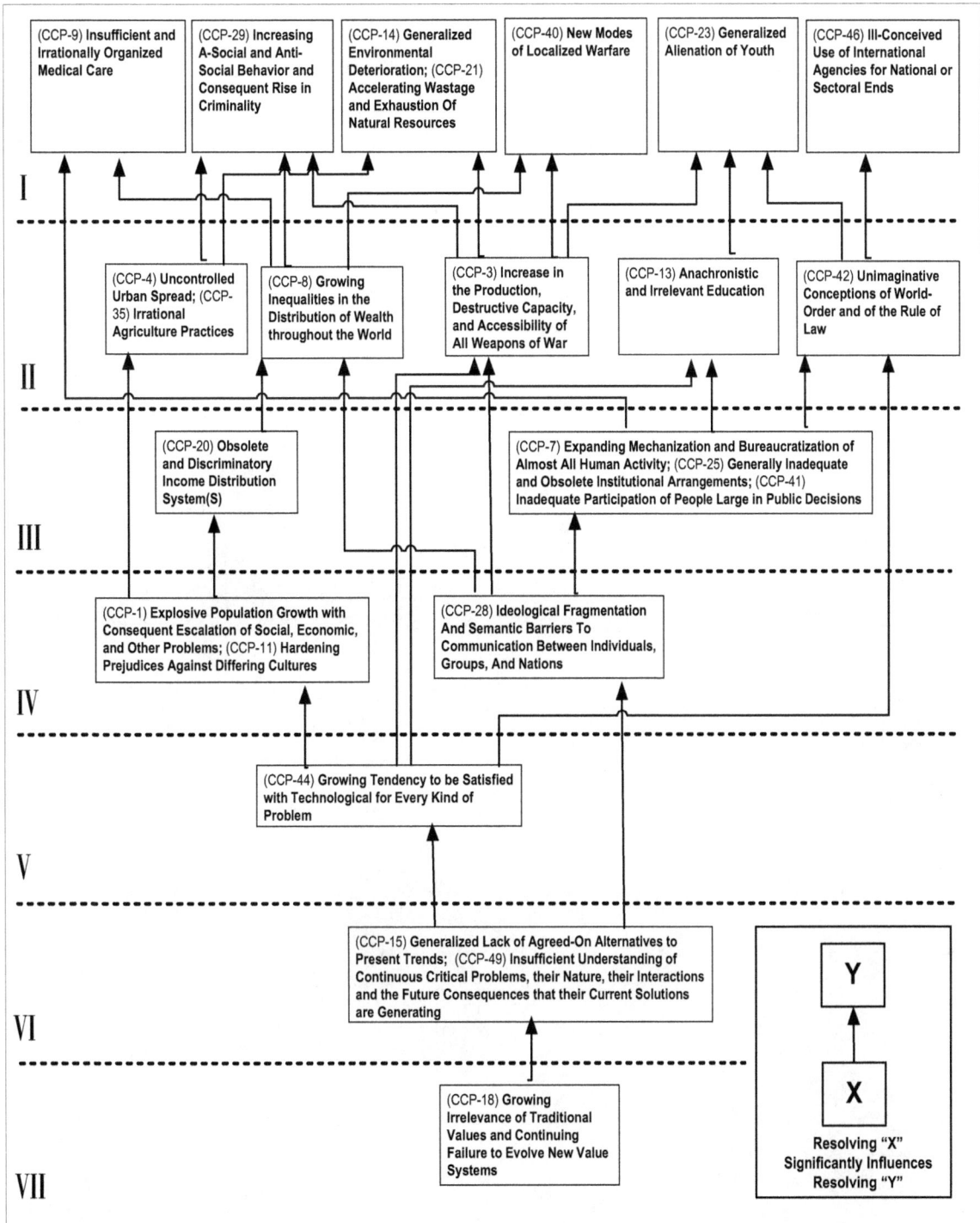

**I**

(CCP-9) Insufficient and Irrationally Organized Medical Care

(CCP-29) Increasing A-Social and Anti-Social Behavior and Consequent Rise in Criminality

(CCP-14) Generalized Environmental Deterioration; (CCP-21) Accelerating Wastage and Exhaustion Of Natural Resources

(CCP-40) New Modes of Localized Warfare

(CCP-23) Generalized Alienation of Youth

(CCP-46) Ill-Conceived Use of International Agencies for National or Sectoral Ends

**II**

(CCP-4) Uncontrolled Urban Spread; (CCP-35) Irrational Agriculture Practices

(CCP-8) Growing Inequalities in the Distribution of Wealth throughout the World

(CCP-3) Increase in the Production, Destructive Capacity, and Accessibility of All Weapons of War

(CCP-13) Anachronistic and Irrelevant Education

(CCP-42) Unimaginative Conceptions of World-Order and of the Rule of Law

**III**

(CCP-20) Obsolete and Discriminatory Income Distribution System(S)

(CCP-7) Expanding Mechanization and Bureaucratization of Almost All Human Activity; (CCP-25) Generally Inadequate and Obsolete Institutional Arrangements; (CCP-41) Inadequate Participation of People Large in Public Decisions

**IV**

(CCP-1) Explosive Population Growth with Consequent Escalation of Social, Economic, and Other Problems; (CCP-11) Hardening Prejudices Against Differing Cultures

(CCP-28) Ideological Fragmentation And Semantic Barriers To Communication Between Individuals, Groups, And Nations

**V**

(CCP-44) Growing Tendency to be Satisfied with Technological for Every Kind of Problem

**VI**

(CCP-15) Generalized Lack of Agreed-On Alternatives to Present Trends; (CCP-49) Insufficient Understanding of Continuous Critical Problems, their Nature, their Interactions and the Future Consequences that their Current Solutions are Generating

**VII**

(CCP-18) Growing Irrelevance of Traditional Values and Continuing Failure to Evolve New Value Systems

Y

X

Resolving "X" Significantly Influences Resolving "Y"

31

**Chapter 3**

**Understanding Structural Inquiry as a Design Process**

This section of the workbook talks about structured dialogic design as a specific tool for managing structural inquiry. It begins with a brief consideration of the philosophical foundation for collaboratively solving complex problems. Structural inquiry can be used, however, by both individuals and groups. We will provide instructions for using tools which support the practice of Structured Dialogic Design® (SDD) as a means for building individual understandings in this section of the workbook. The subsequent section will deal with using the tool in groups for jointly discovering a shared understanding of complex problems and for collaboratively designing action plans to address those problems.

Both this section and the following section of this workbook are intended to be used with CogniScope software which automates and supports Structured Dialogic Design. This software is either included with this text or may be downloaded without cost from the website of the Institute for 21st Century Agoras. Currently the software runs only on Windows®-compatible computers.

PROBLEM SOLVING IS SPONTANEOUS – SHARING KNOWLEDGE IS NOT

Interactive Management (and its evolved form, Structured Dialogic Design) is based on natural ways of making decisions. In response to a focused question, decision makers gather information, organize the information so that alternatives are explicitly recognized, apply some process for making a choice, identify and use the chosen outcome, and (usually) consider how effective the process seems to have been for the decision making task. In short form, decision makers consider alternatives, make choices, and then use those choices. The decision making process matters because the extent to which decisions are put into practice depends upon the perceived rigor with which alternatives are considered and choices are made. Where risk is high, formal and "proven" design and decision making processes are desired.

High risk generally means that the decision makers will be putting significant resources at risk in a venture for which there can be personal victories or defeats. Of course, as you read this you may be saying "Wait a minute! We do this all the time … without any formal processes whatsoever!" You might be saying that even when we are making some of the most important relationships that we will have in our lives, we consider the alternatives and then make our choices based upon our intuitions.

OK, yes, we are not likely to use factor analysis or analytic hierarchy process for such personally critical life choices … we probably can live happy lives without ever knowing about such tools unless we are in some corporate or government risk management team. Our individual tendency is to use and rely upon our intuition – our "body wisdom" – when we are confronted with complexity that doesn't lend itself to addition and subtraction.

Some business managers celebrate individual body wisdom as their "gut feeling." We feel – with our entire body, not only our gut -- that we know the "right" choice. Through the strength of our feelings, we act on that choice.

Reliance on body wisdom gets tricky when we cannot make entirely independent choices or actions. Under these situations, we need to either work with someone who happily has the same gut feeling that we do, or we need to explain our understanding in a way that will resonate with the body wisdom of others.

In the world around us, there are individuals who will happen to share our feelings about choices and others who will not. As a species our history is marked by our habit of finding and working only with those who agree with us and shunning or attacking those who do not. This tradition might be said to have carried us to where we are today – into a world with environmental suffering on a global scale, social and financial inequities worldwide, and violent clashes on basic beliefs.

Sustainability is not simply a matter of making changes, but more fundamentally it is a matter of making changes together. To act together and make changes together – and specifically when such changes affect personal behaviors and lifestyles – we need to share the deep feeling that the proposed changes are necessary and appropriate..

---

**EXERCISE: Consider the statement "Voting is usually a tool for delegation and rarely a tool for direct action"**

A local or regional sustainability event is designed by sustainability experts. The event attracts citizens interested in sustainability. Many, many problems and many alternatives for action are presented, as are inspirational success stories from other communities. The audience seems to like some alternatives more than others. A vote is taken as an act of audience participation and one alternative has received the most votes. Sustainability event leaders declare the preferred alternative as the appropriate alternative for action.

The audience recognizes that a "fair vote" was taken. The event closes with stated goals of acting on the preferred alternative. The participants go home and share the outcome of the vote with friends and family.

Participants are asked "why was it chosen?" They say "it got the most votes!" Participants are further asked "why did it get the most votes?" They say "many other alternatives got fewer votes." Participants are then asked "were the other alternatives better choices?" They say "I don't really know because we didn't actually compare the strengths or weaknesses of alternatives in a side-by-side fashion."

Participants are asked "are you going to invest your energies in the selected alternative?" They say "I am not sure. It probably depends upon whether a steering committee can talk enough of us into that specific action."

So here is the issue. The well-intended event consumed citizen time and energy, and also filled that place on the civic calendar reserved for "discuss sustainability." Citizens of good will attended. Many, many options were presented. Choices were made. One alternative was given legitimacy for action through a popular vote. BASED UPON YOUR EXPERIENCED VIEW OF THE WORLD AND RESPONDING WITH YOUR OWN BODY WISDOM, WILL EFFECTIVE ACTION BE TAKEN AND SUSTAINED?

Do you have the tools for explaining your body wisdom on this situation to others?

PROBLEM SOLVING NEEDS TO BE DEMOCRATIC

The logic which underlies our capacity to solve problems is based upon our assessments of influences between cause and affect. We can disagree in a thousand ways, and yet we need to come to agree in a common way. We might look at events from different perspectives and differ on their possible causes or combinations of causes. We might disagree on the strength of causes or even of the effects themselves. We do not deny cause and effect in our lives; but to let go of one causal belief, we do need to understand (which means to logically accept) the possibility of an alternative causal link.

Our minds can tell us that a causal link can exist, but only our body wisdom will tell us if we feel (based on the balance of our life experiences) that a specific link is as strong as or stronger than an alternative link. By sharing "stories" (accounts) that relate to our strong feelings, we can understand each others' perspectives of cause and affect, and through this understanding we may be able to point to observations which can lead individuals to change their understandings. This process is both as simple and as profound as the essential human skill of "learning." Individual learning and group learning, however, have different prerequisites. Individual learning can have its internal prejudices, but the internal learning community within an individual's own mind shares reflections in an instinctively democratic fashion. As a matter of behavioral health, your inner voices speak in the same volume, yield to each other in turn, and play by the same set of logical rules. Your multiple perspectives are not preoccupied with defeating each other but rather with achieving some sense of "truth" together.

## PROBLEM SOLVING NEEDS TO BE LOGICALLY COHERENT

Internally, our democratic discussions are coherent, that is, they apply "rules" of logic in a fair and consistent fashion. The human mind is hardwired to apply three types of logical approaches for predicting or inferring outcomes. You don't have to consciously "choose" which approach to apply, but rather your mind pulls up the tool automatically and you compare your experiences systematically using these tools. In a very brief fashion:

1. To make an inference (a prediction) using deductive reasoning, your mind reaches back to a situation and links that situation to something that you have experienced as a relevant "precondition." *"When my little brother didn't have a good breakfast yesterday, he was very cranky. He missed his breakfast this morning, so he is probably very cranky now."*

2. To make an inference using inductive reasoning, your mind seeks to create a rule based on sets of parallel experiences and then uses that rule to support deductive reasoning. *"My little brother complains if his breakfast doesn't have something hot and he complains if his breakfast doesn't have something sweet, so a good breakfast must have both of these things. Since he didn't have his special sweet cereal this morning, he will be cranky today."*

3. To make an inference using abductive (or retroductive) reasoning, your mind considers both a conclusion and a rule and then "back calculates" to infer a prior condition. *"My little brother is cranky when he has not had a good breakfast. He is very cranky now, so he must have missed his breakfast this morning."*

When you share an idea for a cause and effect relationship, you share underlying meaning too. You share an understanding of the meaning in relationship to other meanings. This convergence of shared meaning brings perspectives together. At some level, the amount of shared meaning brings perspectives close enough to evoke a feeling of intimacy. Your body wisdom allows you to sense when this convergence is happening. Channels of trust are then opened – incrementally – and meanings that are closer to individuals' deepest hopes, fears, dreams, and dreads surface and become discussable. Only rarely will complete harmony be achieved, but most often democratic dialogue will get distinct perspectives "close enough." Even in an exclusively internal dialogue, the "closure" that one feels when a decision is emerging can be felt as a comforting ripple in body wisdom.

## PROBLEM SOLVING REQUIRES WORKING WITHIN LOGICAL FRAMEWORKS

Problem situations are discussed – even within our own individual minds – in terms of how issues impact other issues. We reflect upon issues, preconditions for issues, and rules which define preconditions, and beliefs about causes of issues. In this sense, we rapidly establish how an issue is "fixed" relative to other significant issues in our experience. We come to "know" the issue – based upon our contextual knowledge of the way that the world works. In this sense, an issue is placed into a structure. It is part of a framework of understandings. To discuss ALL of the contextual ways in which we have come to "know" an issue can be very difficult. We might have to explain our entire life story (well, maybe not that much, but you get the idea).

When we have to share our understanding with someone who may have limited experiences in common with us, we should expect to discuss how we reached our conclusions. To help others follow our explanation, we may need to provide a physical map of the relationships that we believe exist. We won't have to worry about formally defending our use of inductive, abductive, and deductive logic, but at specific points we may need to offer some explanation of how we "thought about" some specific relationships within our maps. In this first cycle of sharing we simply need to put the big picture of connections onto the table. We need a map.

Or maybe we do not need a map. We may never need to worry about putting any explanation onto the table if we don't plan to share our thinking with anyone else. This section of the sustainability class workbook introduces a way of thinking with a software tool that also generates an easily read map of our logic – for helping us review our own thinking and then also for explaining things to others should that need arise.

INTERPRETIVE STRUCTURAL MODELING BUILDS MAPS OF LOGICAL FRAMEWORKS

Logical maps are easy to build. If components are represented by circles, then the ways that they interact can be represented by lines that connect or link circles. With enough connections, a pattern emerges. The pattern represents our understanding of a situation. There, it is that simple. We need (only) recognize the important components, and link them with the appropriate influence relations. This is the essence of powerful tools such as Systems Dynamics and Concept Maps. It gets messy in the real world, of course. Do you "identify issues" before you "understand issues" or is it the other way around? Working with the speed of an individual mind, you can only identify those issues which you understand, so the relationship collapses – regardless of its importance. A simple logical framework might consist of something like:

1) define a search for relevant issues;

2) construct a catalogue of issues;

3) identify highly preferred issues;

4) bring the most preferred issue onto the map first; and

5) repeat the process with the next most preferred issue that can be linked into the map.

There are some problems with this approach that are worth considering. Yes, you may have missed or even misunderstood some issues. Yes, you may have some erroneous priorities when you ranked the relative importance of issues. And, yes again, you most certainly may have defined links inappropriately or inaccurately. Hey, this is the human condition. Habits of the human mind must be recognized and accepted. There is only so much that one mind can do.

Before moving beyond the simple, back-of-the-envelope approach to constructing a logical map, there is one additional subtlety of the easy approach to consider. When manually mapping a framework, some forms of early choices can shape options for future choices. This is called a design "trajectory." For example, if you continue to construct your list of issues as you are building the map, you can be led to "focus" on the early structure of the map before you have captured the broader landscape. It is natural to focus this way. The more that we think about a certain thing, the less mental energy we have to spend thinking about other things. It is sort of like an economics of logic. The CogniScope® software can speed up your construction of maps and protect you from the pitfall of erroneous design trajectories. Of course the overall structure can be balanced out through iterative refinement of your efforts to adjust a starting point bias, but you may have led yourself down a nonproductive tunnel of thought and spent more time than you otherwise would have to spend. The back-of-the-envelope approach is convenient, but it may also be less efficient for you.

## THE PROBLEM OF ERRONEOUS DESIGN TRAJECTORIES

The CogniScope® software avoids erroneous design trajectories by separating decisions about the relationship between pairs of issues from the actual mapping of those relationships. The software asks you to make a comparative judgment and then it tracks your answer. You remain free to introduce new issues at any time during the process, but you are not steered toward choices based upon preliminary work. For this reason, when you use the software, you will go through a series of issues naming, issues clarification, issues comparison, and then pair wise relationship assessment tasks before you are presented with the structure that you have mapped. The technology that allows the software to do this is discussed in recent books by the software inventor and coauthors (Christakis and Bausch, 2007; Flanagan and Christakis, 2010).

**Chapter 4**

**Using Structured Inquiry for Individual Systems Thinking**

*STARTING WITH **CogniScope® II** SOFTWARE*

*Download and install* the student academic-version of the CogniScope II software (CS II). Go to www.globalagoras.org/CSII/sustainablefuturesstudent. After you enter your name, email address, your academic institution, you can download the CSII software at no cost. Install this software in your computer.

The software is designed for a Windows®-compatible computer and may not run on a Macintosh® computer. Choose "Begin" to start the installation process. You will be prompted to select a location for the files ("C:\Program Files\CS is the default location). You may select another location if you wish. Once the file transfer has been completed, Click "Exit" to leave the install program. You may begin using the CS II software by clicking the desktop icon (if installed) or the Start Menu entry.

This version of the CogniScope software is called "RootCause Mapping." It was manufactured by the Leading Design International (LDI) software company and it was donated as a gift to the Institute for 21$^{st}$ Century Agoras to advance education missions in fields of applied democracy. Your sustainability course is just such an application. The software does not require any user password; however, your use of the software carries the inferred agreement that the manufacturers and distributors of the software are exempt from any and all liabilities with respect to your use of the software. Among other things, this means that if you get a B and had expected an A in this course, you cannot blame the software. Seriously though, there are conditions that can cause some design software to "crash" or "lock up" and for this reason we encourage you to be precise when you follow the step-wise process of using this product. CSII is a solid product, but it has locked up too.

[ **Note:** the minimal steps that you **NEED** to follow are stated in **bold type.** The additional steps provide background for your understanding of the software functions and will help you in your own creative uses of the design tool.]

*OPENING AN EXISTING PROJECT*

When the CS II software is opened, the user is presented with a choice of starting a new project or opening an existing project … and, of course, the option of exiting the software. To simplify your first-use experience we are starting everyone off with a working file that we have prepared for you. **Click on "open an existing project", scan through the pop up window, and then select the file named "INDIVIDUAL".** If you select "New Project" you will begin setting up a new project file and you will not get specific guidance from this workbook. You can explore as you choose, but you must not expect to get new project help in this course. You can find help, however in *How People Harness their Collective Wisdom and Power* and also *The Talking Point.* (See references at the end of this book.)

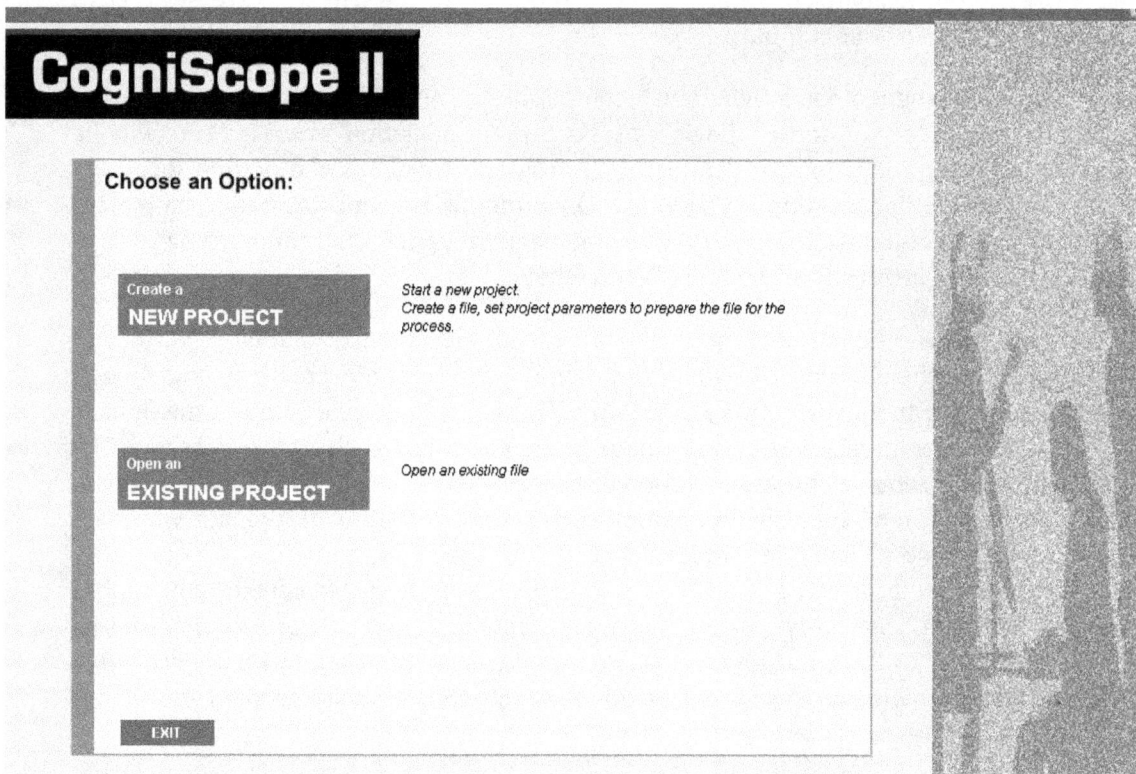

FIGURE 1

Once you have opened this project file, you will be asked whether you are working on a DIAGNOSIS or a DESIGN project. **You should select DIAGNOSIS (see below).** The DESIGN

40

option is for designing solutions once situations have been diagnosed and is thus an advanced function (which will not be discussed in this brief introduction).

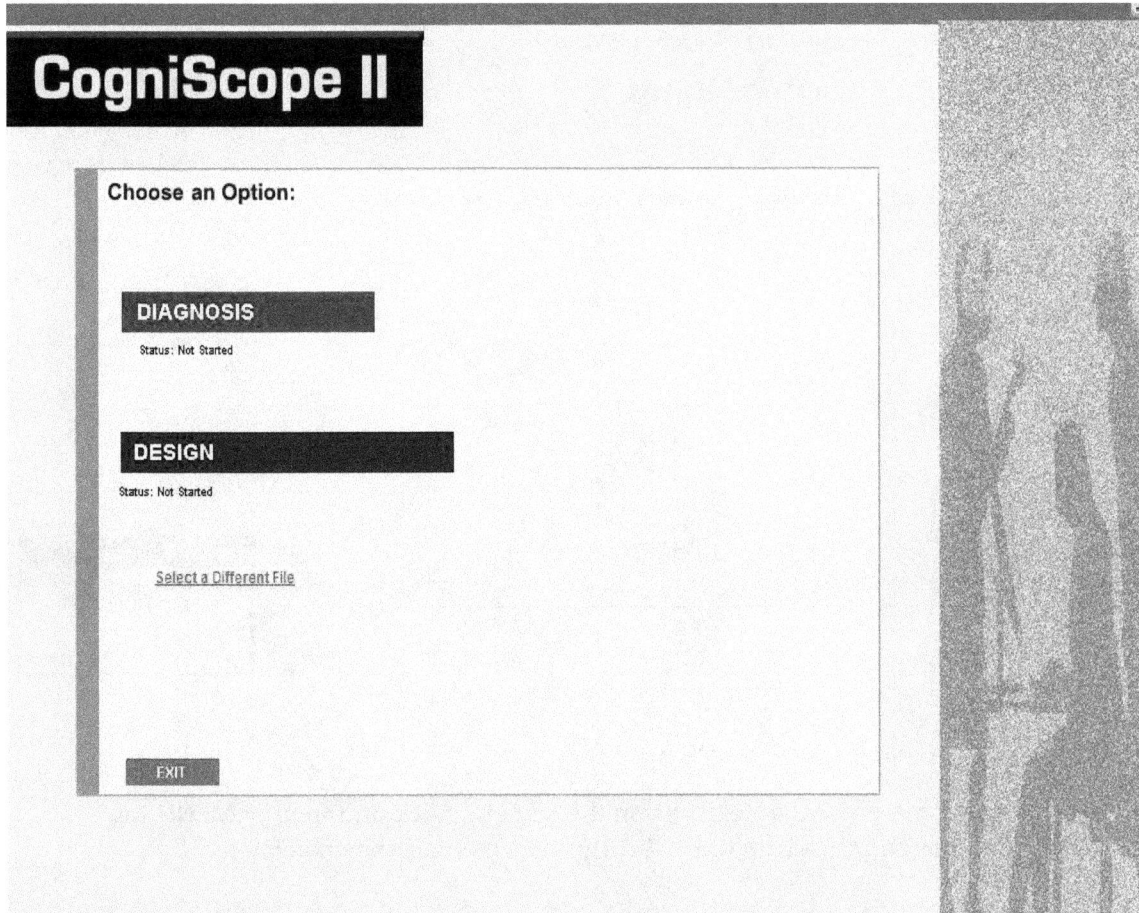

FIGURE 2

**The DIAGNOSIS button will bring you to the GENERATION function. Here you will see a dialogue box with the "trigger question" for beginning the diagnosis.** You already know something about this situation because it has existed a long time and overwhelmed the founders of the Club of Rome when they first considered engaging this question.

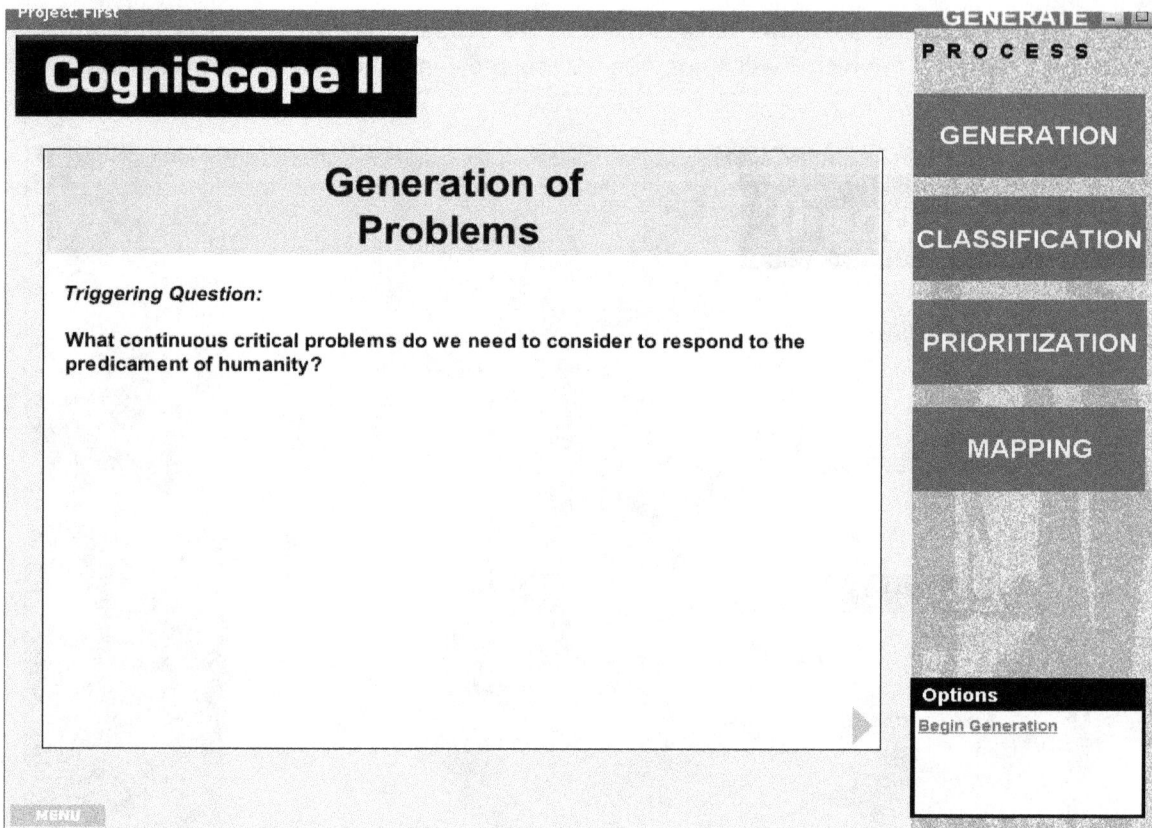

FIGURE 3

Now you will see some unfamiliar features on the screen. **Click on the grey MENU tab at the bottom of the screen to pull up the TABs:** This new menu is shown below:

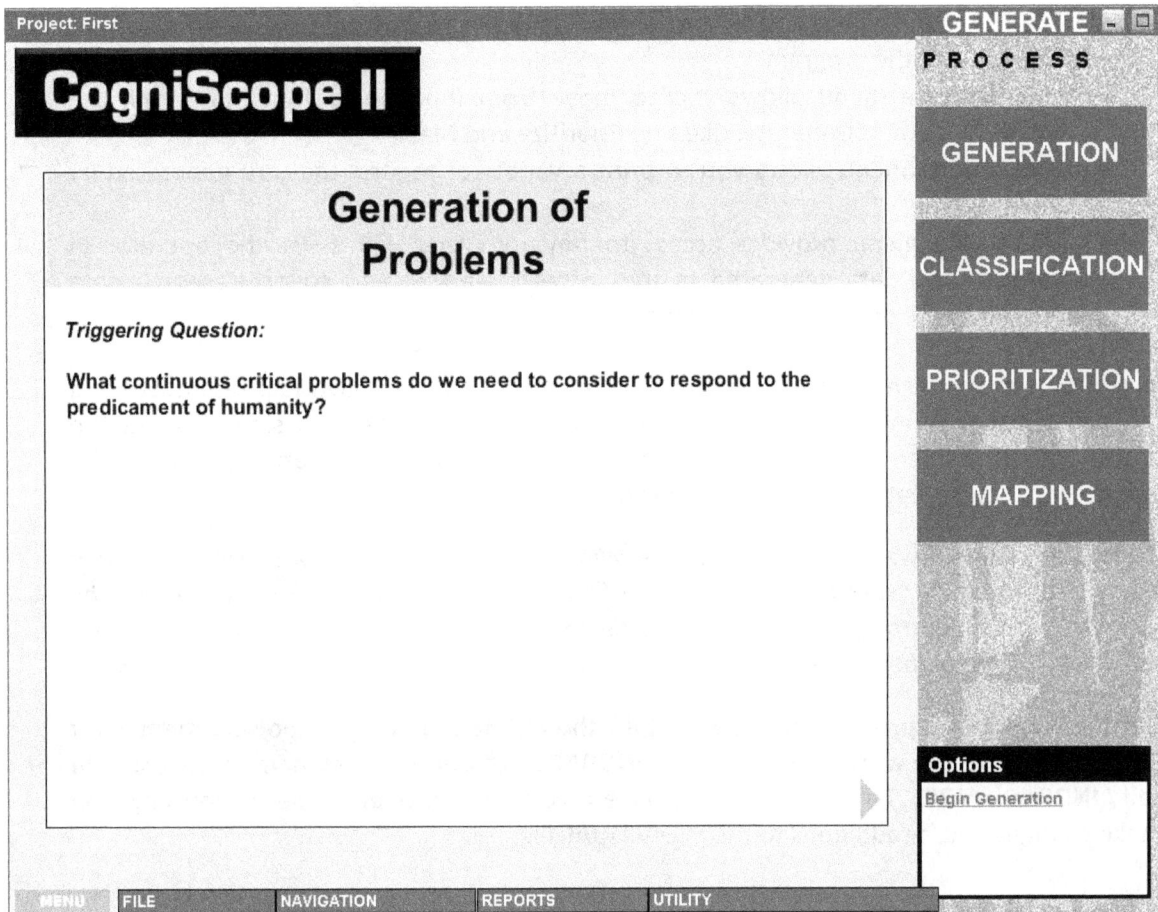

FIGURE 4.

You must click on the MENU tab to get the menu to pop up for you.

*REVIEWING THE SCREEN LAYOUT*

You won't be using this software in its full design mode in this workshop class (though you are free to explore on your own if you choose). For this reason, we will be brief. The design of the CS II Mapping software follows a pattern of having a heading at the top (to let you know the project name and location), a content area in the center of the screen (where you are "working"), and a ribbon of navigation functions located at the bottom of the screen (to jump to other parts of the program). To view the Tabs, press the "Menu" button located in the lower left hand corner of the screen.

- The **File** menu allows you to start a new file, open an existing file, save the current file, and exit the program.
- The **Navigate** menu allows you to move from function to function within the software, such as Generate, Classify, Prioritize and Map.
- The **Reports** menu allows you to print a variety of reports that are generated by the program
- The **Utility** menu provides access to functions that will assist the operator in handling the data generated in the software and also to collect research data related to the decision making process.

In addition to the menu at the bottom of the screen (which we have just discussed) you can navigate from function to function using the buttons on the right side also. At the bottom of the column on the right side you'll see "Options" that are appropriate for the particular function that is currently being used.

(Note from the software designer: *"If the Windows Taskbar covers your menu you can change the characteristics of the Taskbar to "Auto Hide". 1) Right-click anywhere on the taskbar (not on a button). 2) Choose "Properties' from the pop-up menu. 3) On the Taskbar tab, click Auto Hide the Taskbar."*)

**Getting back to our project: Click menu, Click the FILE tab, and in the pop-up menu click SAVE AS to save the working file named "INDIVIDUAL" under a new name that you will call "INDIVIDUAL2".** This allows you to have a copy of the original file as you begin to make changes and/or additions to the INDIVIDUAL file.

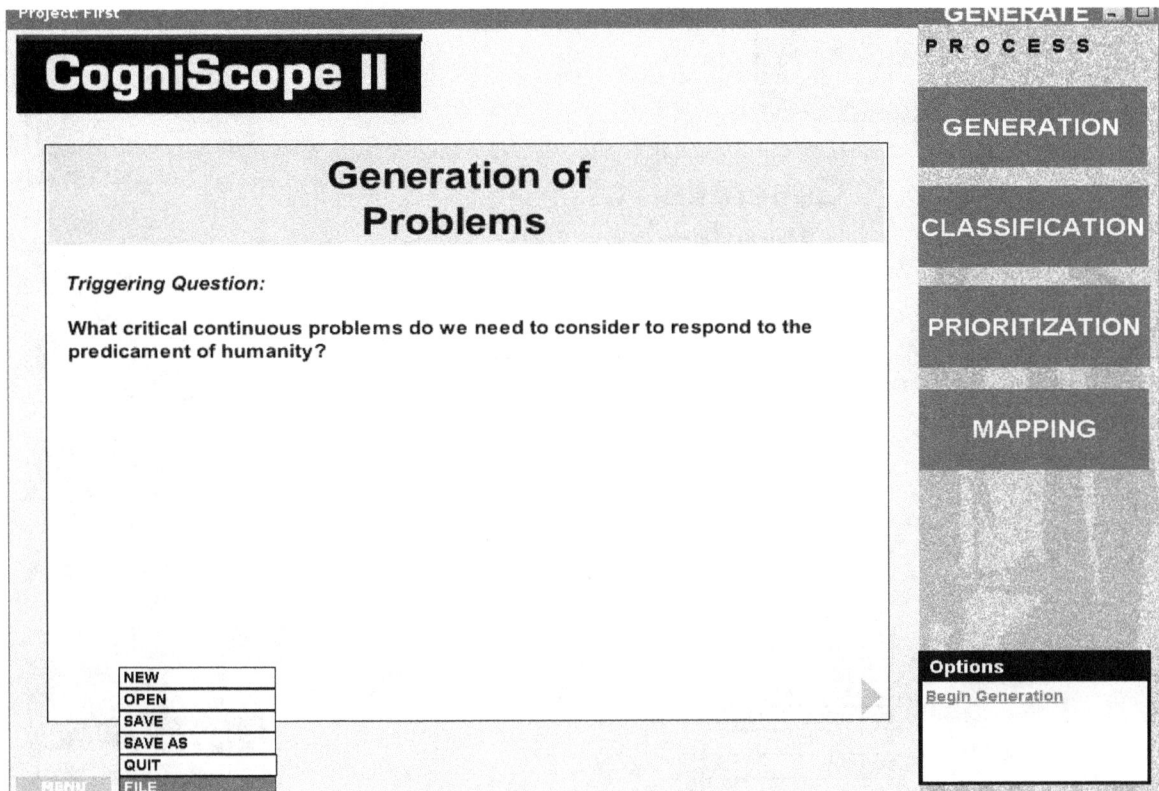

FIGURE 5

Once you have saved a copy of the file that you have named INDIVIDUAL2 for your work, you will be returned to the menu options you were previously viewing. **Click the NAVIGATION tab (as shown in Figure 4) to see the Navigation Menu.**

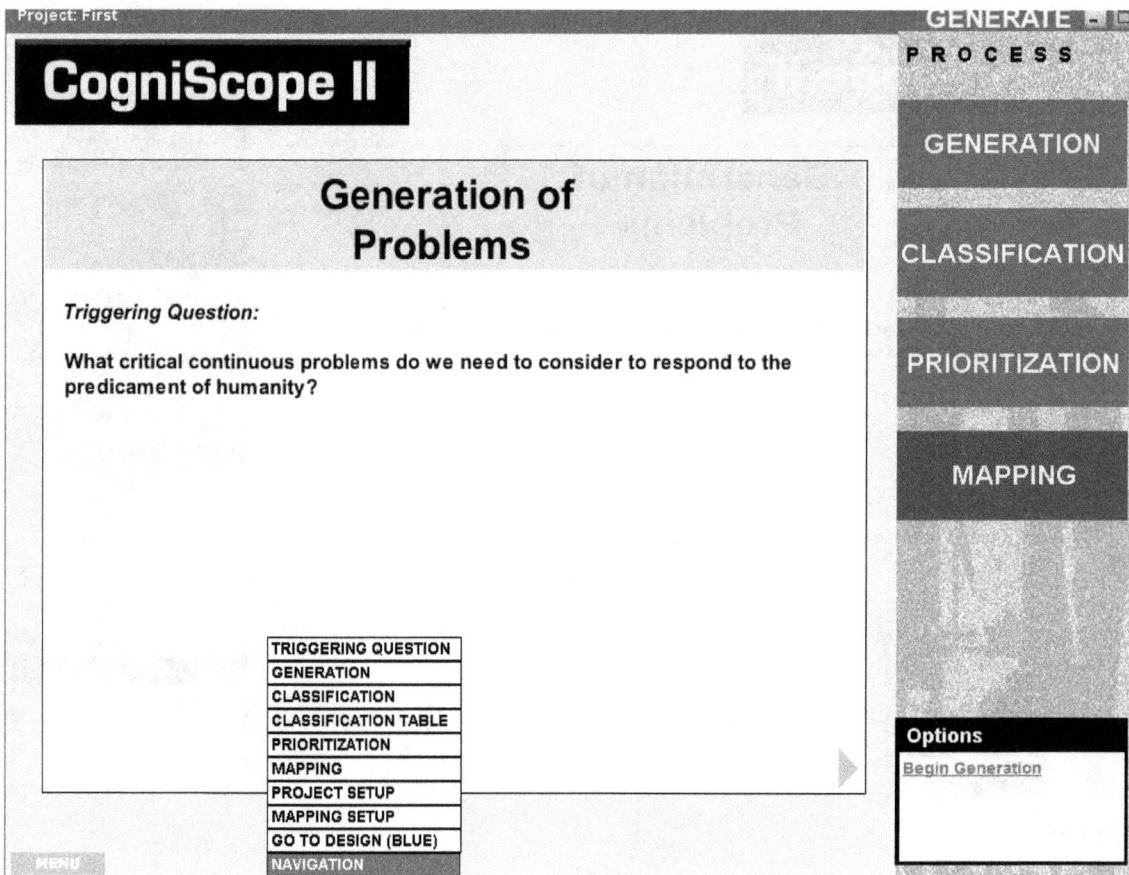

FIGURE 6

In the NAVIGATION tab pop-up menu, you would ordinarily CLICK "BEGIN GENERATION."

In a real-life application of SDD, you would be taken first to a blank screen to begin entering problems. If you were continuing or resuming work on a real-life file, then you would be pulled to the end of the list that that had been created earlier and you would add additional problems to that existing list. **In this chapter, your INDIVIDUAL file -- and now that you have created it, also your INDIVIDUAL2 file -- will have a list of 5 continuous critical problems.** All of your classmates will be working on the same list of five problems at this point. Your personal "INDIVIDUAL2" will differ from other indidivual2 files in the class based on how you assemble the list of problems into a structure. **Your begin reviewing your list of problems in your "INDIVIDUAL2" FILE by clicking "Generation" on the bottom menu**. This command will take you the existing list to review or add new problems – (but you are not being asked to expand the list in this exercise).

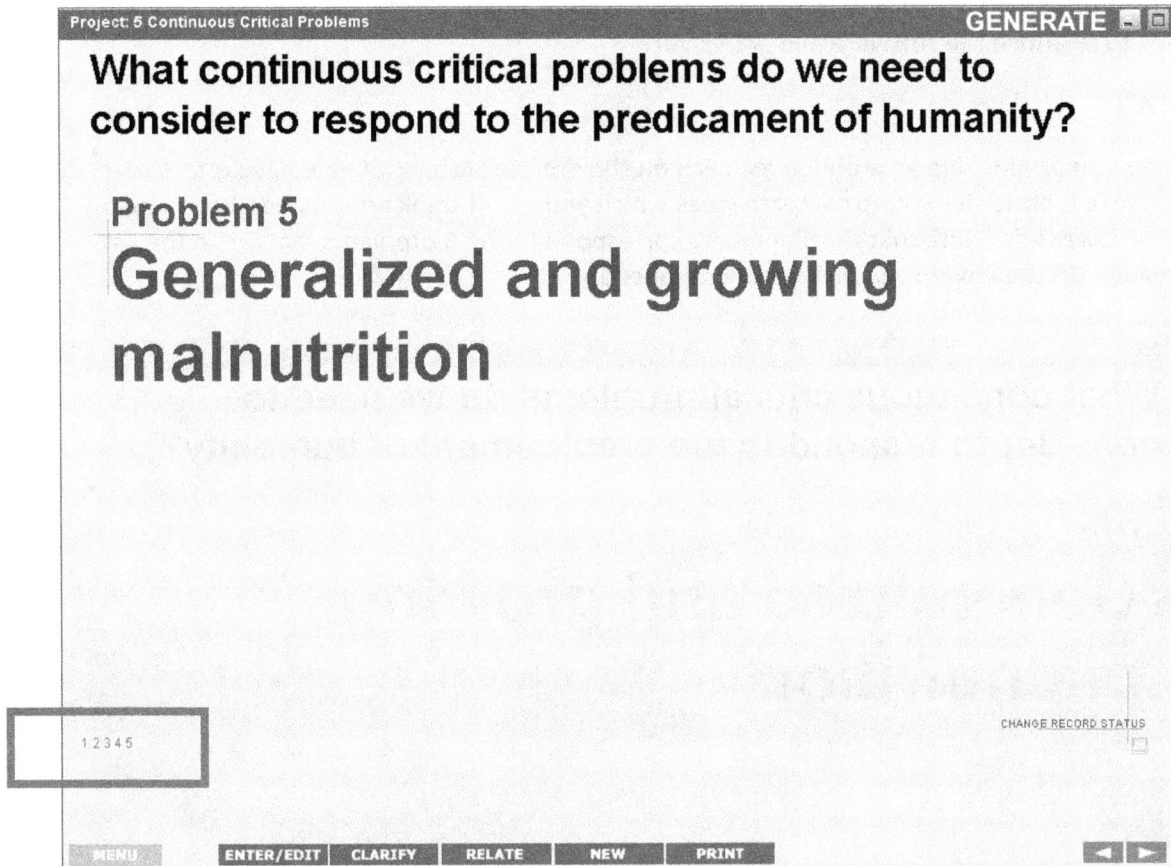

FIGURE 7

In a real-life application of SDD, you would be free to enter additional problems. The classroom exercise in INDIVIDUAL (your "INDIVIDUAL2") file contains only five pre-selected problems. They are CCPs #1 - #5. Figure 7 shows you the view that you will first see when you open "begin generation" for your "INDIVIDUAL2" file. The heavily bordered box will show you that you have five items (problems) in the list for this file.

**Click on the CLARIFY tab.**

**For each of these 5 CCPs you are asked to scan the URLs presented in Chapter 6 to discern the meaning of the problem. Making sure you are in the clarify mode (see Figure 8), type a paragraph explaining each problem into the dialogue box at the bottom of the screen. Yes, you can cut and paste into this box, but** if you try to paste more than text

47

alone – or if you try to paste too much – **your system might *CRASH* on you, and you will have to relaunch the software and start over.**

*RELATE*

**While you are looking at problem #5, click on the "relate" tab**. You now have a tool that will make it easier for you to compare ideas which you might think are related. Drop down to the lower left. Note that the 5 numbers correspond to the 5 problems that are in the list already. **On the lower row, click on the number 1**.

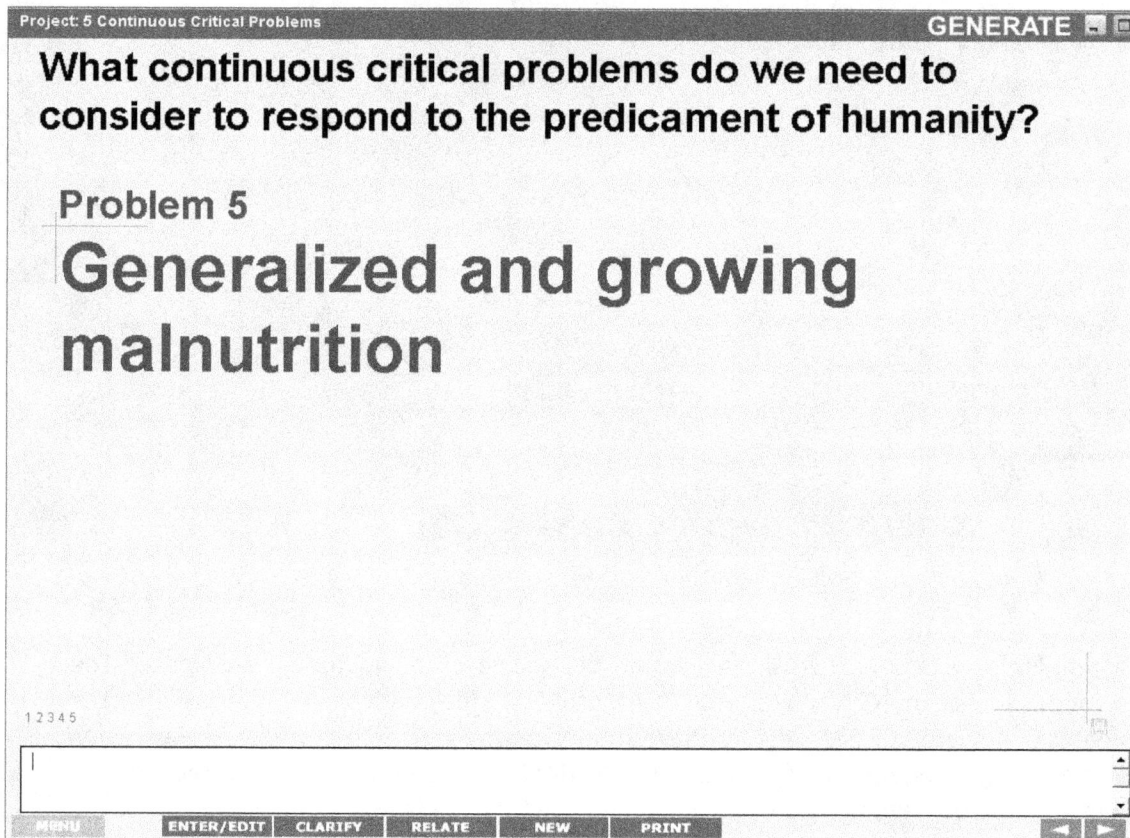

Project: 5 Continuous Critical Problems
GENERATE

# What continuous critical problems do we need to consider to respond to the predicament of humanity?

## Problem 5
# Generalized and growing malnutrition

1 2 3 4 5

MENU    ENTER/EDIT    CLARIFY    RELATE    NEW    PRINT

FIGURE 8

You will now see problem #5 and problem #1 shown together, FIGURE 9. This view allows you to compare two distinct ideas with each other so that you can identify distinctions between them or come to see that they are essentially the same.

We will let you and your classmates sort this type of question out. **How do you sort it out? If you click on the "clarify" tab, you will jump back into the clarifications box for problem**

**#5 and also for problem #1. This will allow you to understand those problems better and decide whether they are mostly the same or distinctly different. Later the split screen will be used to decide what influence these problems may have on each other.** [Your decision is to be made by your reason in harmony with your body wisdom. There are no completely right or completely wrong decisions, but only reasonable ones.] You will see this process played out as you move deeper into this project.

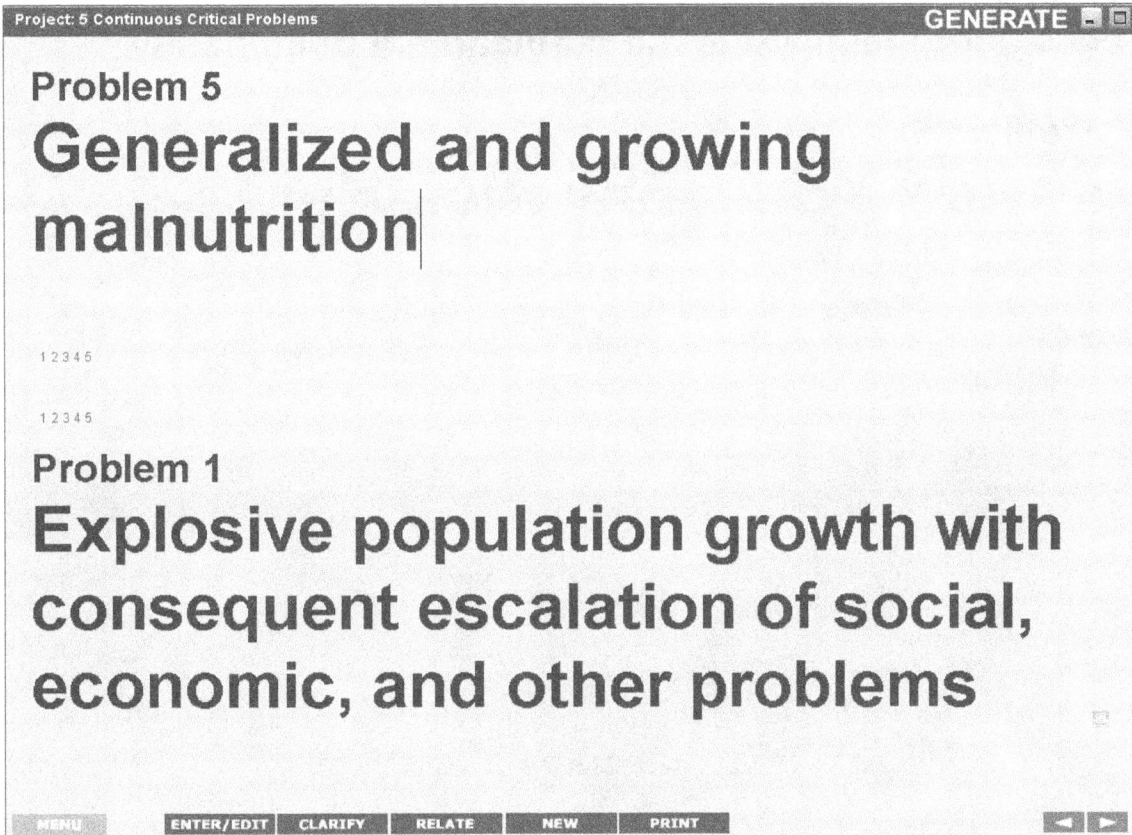

---

Project: 5 Continuous Critical Problems                                      GENERATE ▪ ▢

## Problem 5
# Generalized and growing malnutrition

1 2 3 4 5

1 2 3 4 5

## Problem 1
# Explosive population growth with consequent escalation of social, economic, and other problems

MENU    ENTER/EDIT    CLARIFY    RELATE    NEW    PRINT           ◄ ►

---

FIGURE 9

\*   \*   \*

**[To follow this class, you do not have to learn the next four paragraphs or FIGURE 10]**

Maybe you (or your classmates) have come up with a problem that you feel is critical and continuous and has not been captured in Hasan's list. This is possible. For real life SDD sessions, it is probable. In the event that you feel you should add a new problem, click on "new." You will be prompted with a blank page for your NEW problem (we have typed <<

49

Add Your New Idea Here >> into this blank page just to remind you that you need to put the idea at the top of the page. Once you have put in a label for your NEW idea, you may then enter a clarification for the idea by clicking on the CLARIFY tab. After you do this, it is a good idea to save the file.

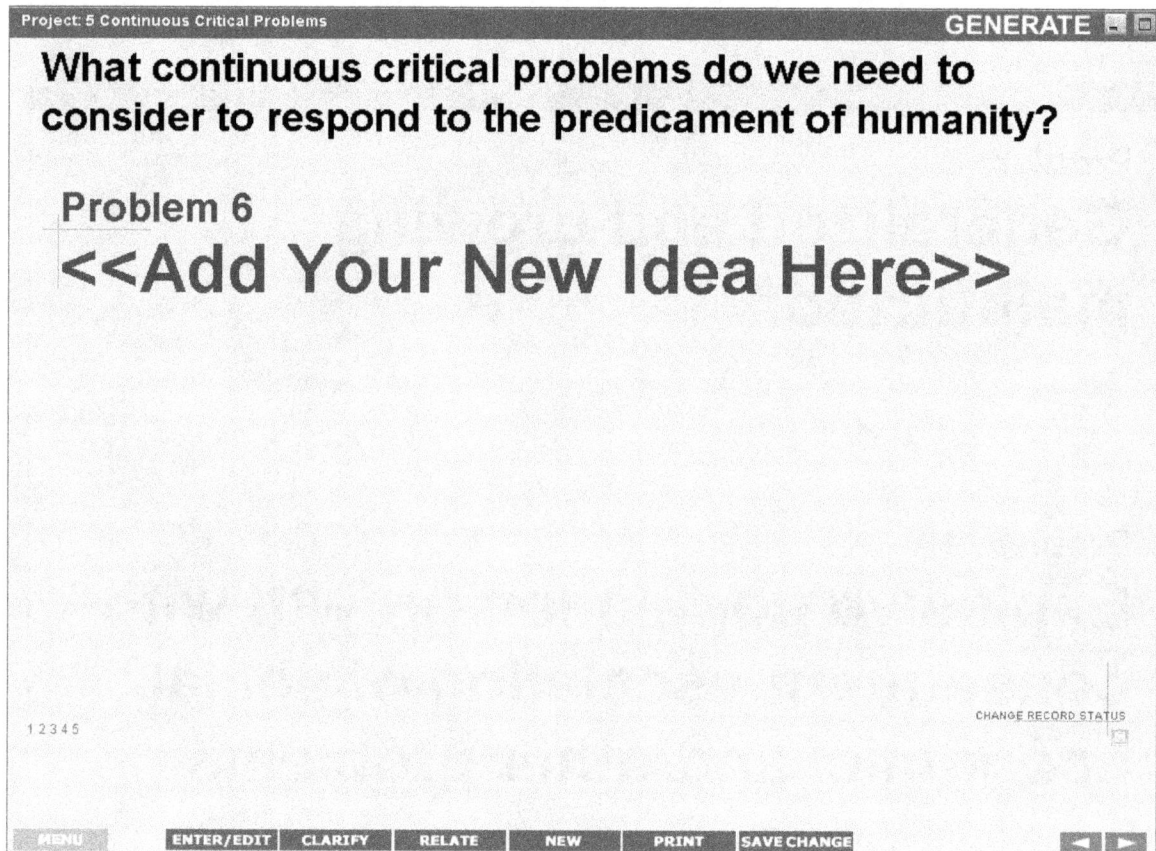

Project: 5 Continuous Critical Problems — GENERATE

## What continuous critical problems do we need to consider to respond to the predicament of humanity?

### Problem 6
# <<Add Your New Idea Here>>

1 2 3 4 5

CHANGE RECORD STATUS

MENU | ENTER/EDIT | CLARIFY | RELATE | NEW | PRINT | SAVE CHANGE

FIGURE 10

*ENTER/EDIT.* You enter the edit mode automatically within some subroutines of CogniScope. GENERATION is one program where you can type into an existing screen directly.

*PRINT.* **You will not need to print from the GENERATION subroutine. However, printing will be automatic if you enter a new field and then click the NEW tab. To avoid printing a page, use the ENTER/EDIT tab first and then click the NEW tab.** The reason that CogniSystem automates printing is to allow group support operations where groups rapidly enter labels of new ideas into a list and the CogniScope team needs to rapidly print up wall displays. This will not be applicable to your individual work, however, your instructor (or a designated student) will need to print up all of the problems in the list that Hasan has created in preparation for your face-to-face group work in the subsequent section (**but in the interests of preserving paper and applying sustainable practices please do not print at this time).**

*SAVE CHANGES*. The SAVE CHANGES tab appears when you create or edit the text in the GENERATION text box in the process of constructing or modifying lists. Once activated, the SAVE CHANGES tab remains open until you click it -- even as you move across items in the GENERATION list. The tab will not appear if you leave the GENERATION subroutine and subsequently return to it.

*   *   *

At this point **you should get back to the screen view shown in FIGURE 3. While there, click on the MENU tab to bring back the NAVIGATION tab, and then select GENERATION from the pop-up menu.**

**You are now ready to work with the list of problems.** You will **enter the CLASSIFICATION subprogram to explore its features and you will perform a quick mechanical classification of the five (5) problems in the list. You will need to consult the clarification section in Chapter 6 of this workbook and you also will need to do a bit of Internet research to understand the problems that you will be structuring.** As you do this task, please remember that the significance of the sorting will mean much more when you engage the complexity of a large list during the class room project (Chapter 5). **This section will have all members of the class individually sorting the first five problems of Hasan's list. Later the group will compare and discuss your individual classification work.**

*CLASSIFICATION*

**Enter the classification subprogram and click on the "begin new classification" prompt.**

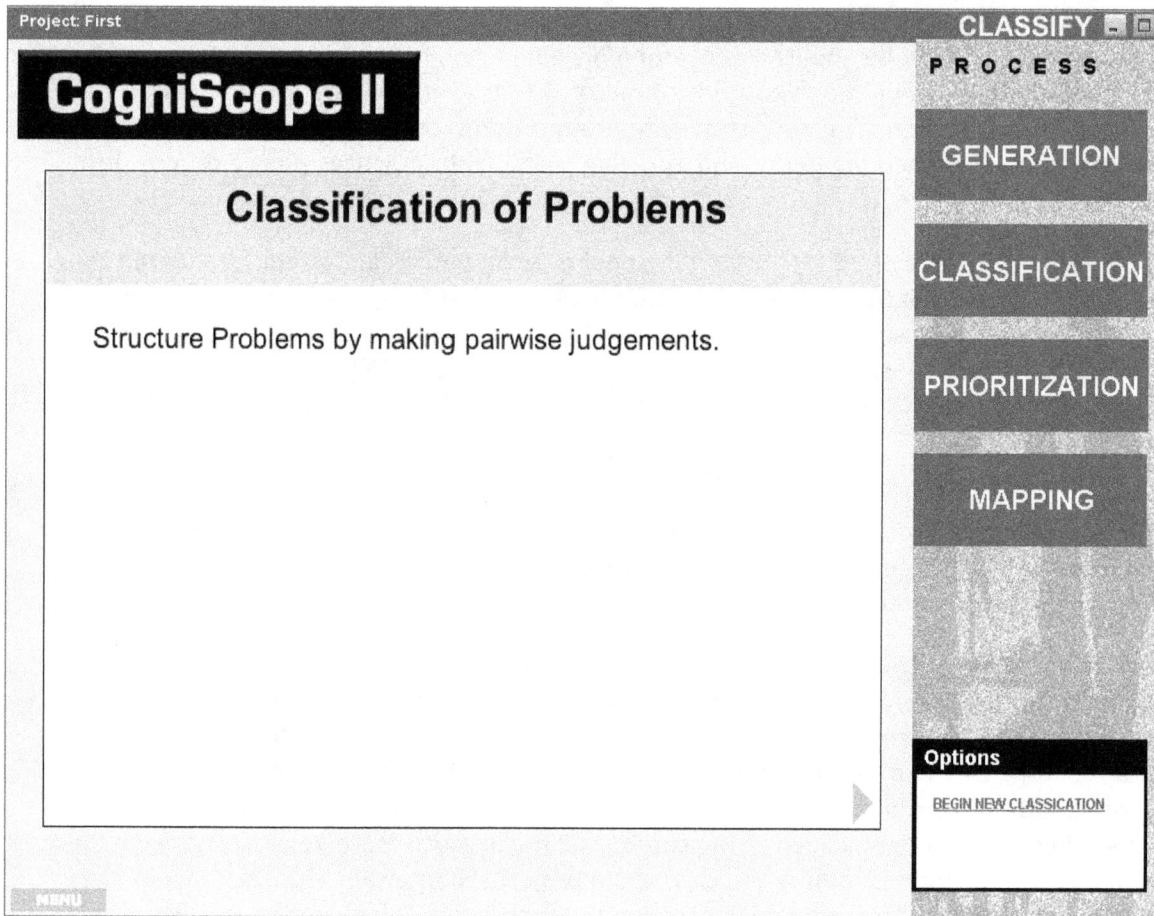

FIGURE 11

The "Begin New Classification" prompt will present you with a screen within which the first problem in the problem list will appear in a window that is labeled as "Cluster 1." You will be prompted to make a judgment. **Does problem #2 "have significant common attributes with" problem #1? This is the question that you will ask yourself as you determine if problem #2 and problem #1 should be put into the same cluster.** Don't worry, you will be able to edit the clusters – and when working as a group, there can be considerable give and take as you discuss the merits of clustering problems into a small set of large clusters or splitting them up into a larger number of smaller clusters.

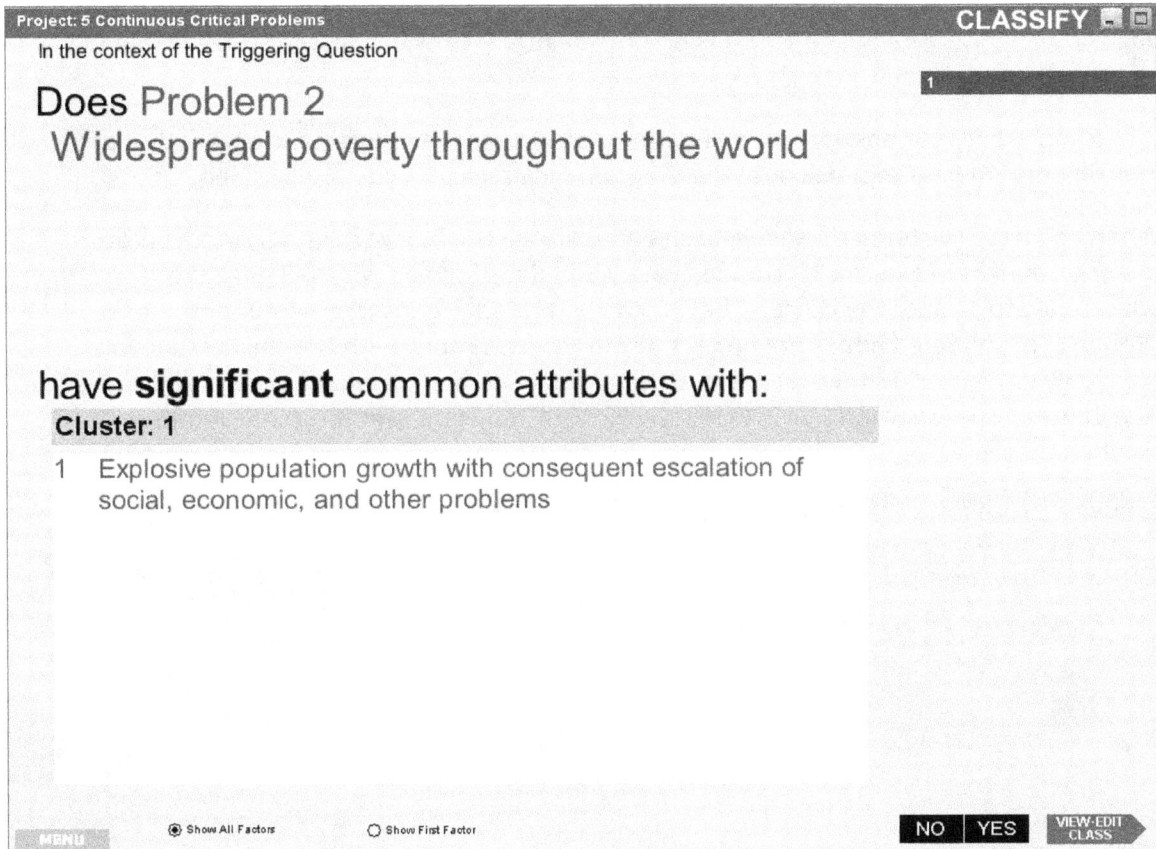

FIGURE 12

You will enter your decision as a "NO" or a "YES" by clicking on the choice button on the lower right of the screen. You also will have a navigator (see FIGURE 3) that will allow you to jump across your set of clusters to rapidly inspect them and see the ideas that they contain.

This decision making process is "abductive." It is an exercise that is seeking to get at your understanding of rules for clustering types of problems in the same bin. You are not being given the name of the cluster because naming the cluster can prejudice you to think in terms of preexisting categories. Here you will be "discovering" the categories by doing the categorization task directly. **As you consider whether "Widespread poverty throughout the world" belongs with "Explosive population growth with consequent escalation of social, economic, and other problems" ask yourself "WHAT ARE THE ATTRIBUTES – OR THE INTERNAL FEATURES – OF THESE TWO TYPES OF PROBLEMS … AND DO THE SETS OF ATTRIBUTES "FEEL" VERY SIMILAR TO YOU." Reflect for a few moments, and then go with your body wisdom.** [When you do this with a group, you are likely to be asked to explain your "feelings" about your selection. When this happens, your logic will be called upon to generate a rationale to explain your feelings. We will discuss this phenomenon more in the next chapter.]

When you enter your answer **you will be prompted to make a judgment on another pair of problems. Repeat this task until the program indicates that you are finished.**

**When you have clustered these five problems, click on the "View / Edit Class" arrow at the bottom right of Figure 12 .** On the "View / Edit Class" screen, you will have the opportunity to "name" the clusters that you have created. **To insert a name for a cluster, click on the numbered box in the upper right that corresponds with the cluster that you are viewing.** Figure 13 shows a case in which problem #4 was placed in a cluster by itself (see Cluster #3). The words "Click to Display Factors" appears over the stack of these numbered boxes.

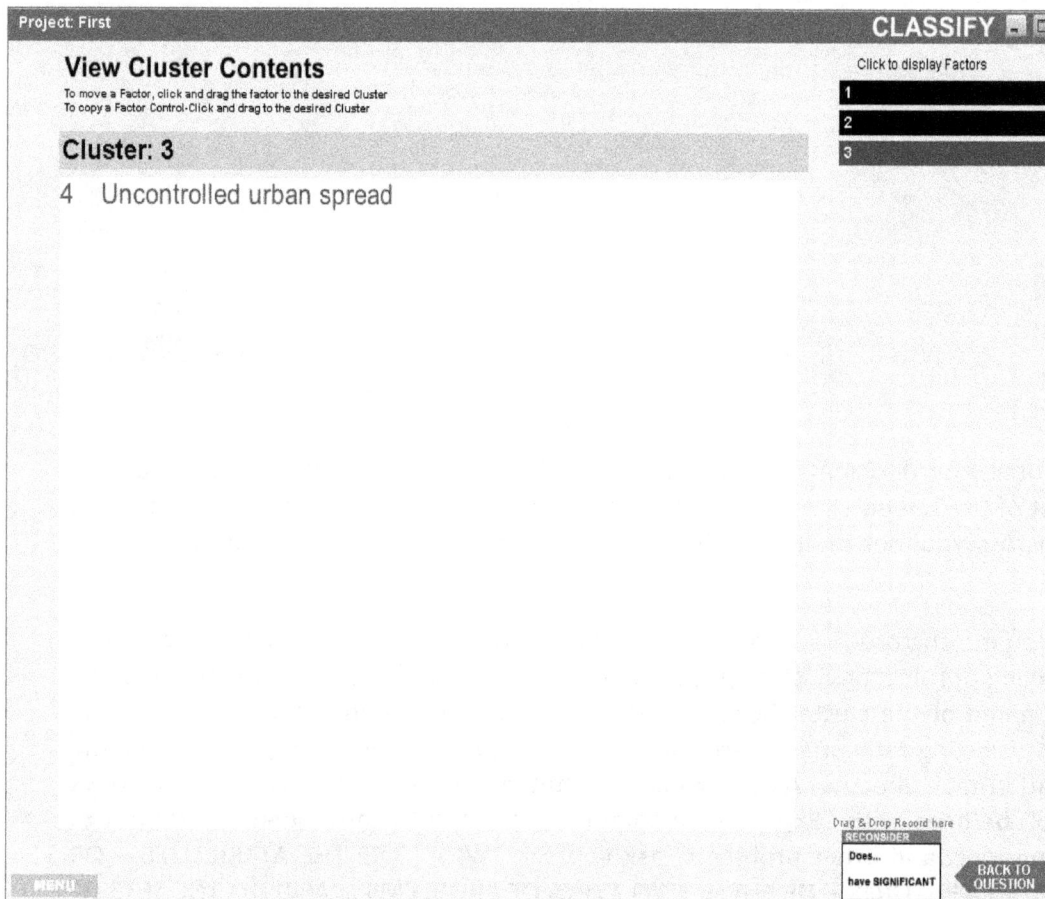

FIGURE 13

**When you "right click" on the highlighted box, a pop up will appear. You can then enter the name you have chosen for this cluster in that pop up box.** In the example shown in Figure 14, the word "sprawl" was typed into the pop up box as a label for Cluster #3. Your clusters will reflect your individual body wisdom and are expected to be a distinct reflection of your individual understanding. Only through group discussion can we expect to approximate what we could agree to be the "real" or the "consensual" clusters. Continue this task to label your clusters. **When you click the word "save" on the pop-up box, the pop-up box will close and your name for the cluster will then appear in the highlighted box on the screen.**

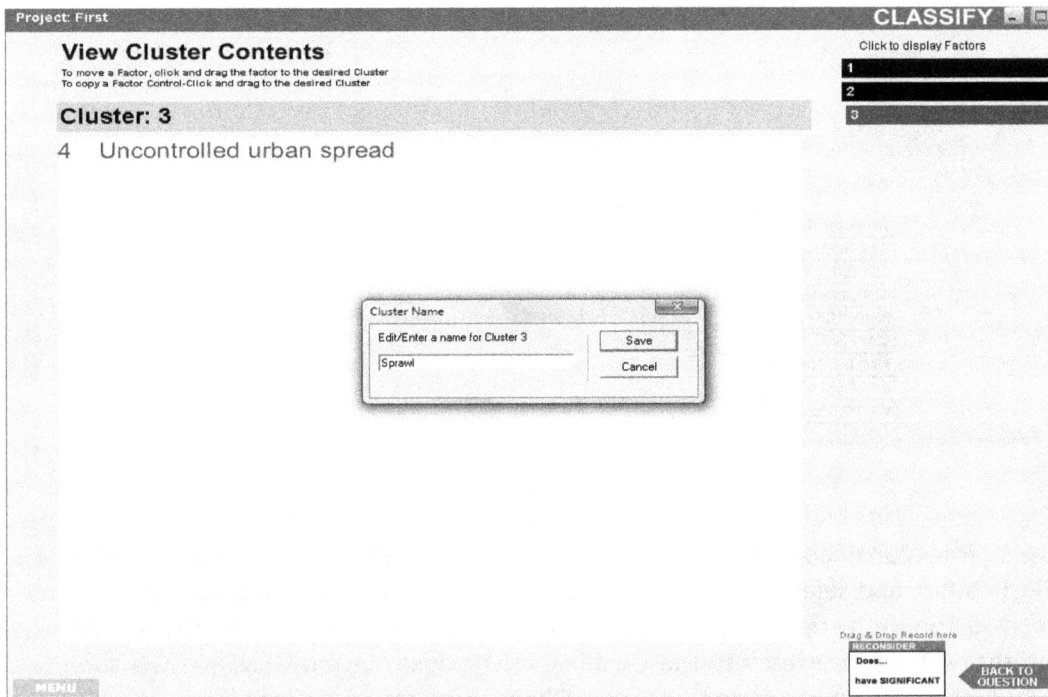

FIGURE 14

Figure 15 shows how cluster names can be entered or changed from the Classification Table view of the data that you have organized.

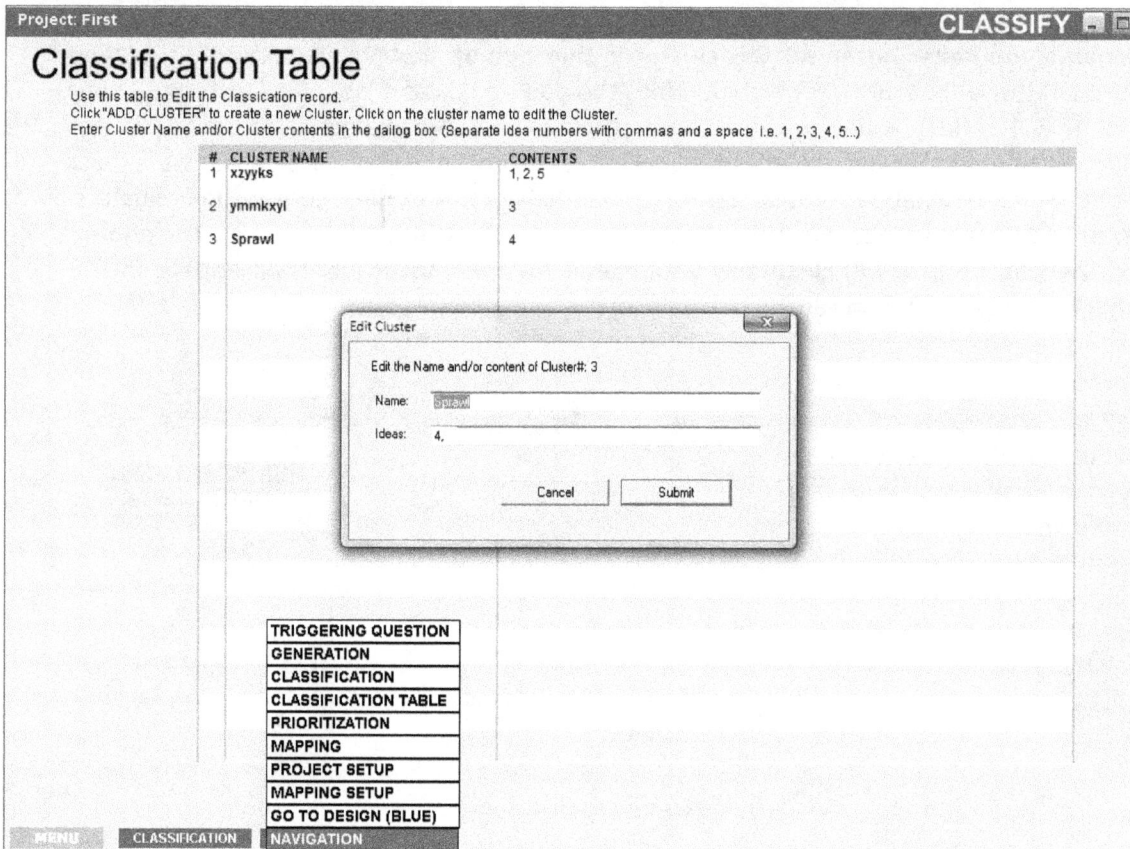

FIGURE 15

You can get a professional looking "report" of your clusters by going to the MENU, clicking on the REPORTS tab, and selecting "Figure 1. Clusters." While you are here, you will also realize that you can get a report for the lists and also for the lists annotated with the clarification that will be inserted into the CLARIFICATION boxes in the GENERATION sub-program. And, of course, you can see that you will have more report options too.

The CLASSIFICATION sub-program is designed to carry a group all the way through classification of the entire set of problems in the list of continuous critical problems. You can explore this program as you choose. In this chapter, we are identifying the minimal tasks that you need to do to understand how the tool works and to begin working with it.

At this point you have gained some knowledge about a tiny bit of the set of 49 continuous critical problems. You now have an understanding of the "dimensions" of the problem situation as the problems are spread into affinity clusters. This is, of course, a tiny bit of new knowledge – and it is observer-dependent knowledge. You know a bit more about how you think about the situation. In a group setting, you will come to understand both how you think about the dimensions of the situation and how the group thinks about them. And in a group, you will have the opportunity to participate in discussions that can shape and reshape the group's thinking and understanding.

PRIORITIZATION

**With this incrementally added understanding of the problem situation, it is now time for you to express some preliminary "preference votes."** The task (for this exercise) is strictly mechanical because we are going to ask you to cast five votes on five distinct problems, and since you have only five problems to consider in your structure at this point (unless you have chosen to go beyond the minimal task we have assigned in this section), you will be placing a vote on each and every problem that you have structured. When you have a larger set of problems, clear preferences will emerge. This is striking in a group process.

**Next, go back to the NAVIGATION tab and select PRIORITIZATION. A prioritization screen will be seen and you will be given the option to "Enter Voting Results."** You will see a table that contains a list of continuous critical problems and a highlighted vertical column with the number "0" in it (Figure 16). Your task is to enter a tally of the votes for the preferred problem ideas into this table. BE GENTLE. You MUST replace the "0" with a number, and you cannot edit this table faster than the software can respond, **so take your time with this task to avoid crashing the software. Replace the "0" in the first five problems with a "1" to reflect your individual vote in preference for the five problems that you have structured.**

**Project: 5 Continuous Critical Problems**      PRIORITIZE

| # | | |
|---|---|---|
| 1 | 1 | Explosive population growth with consequent escalation of social, economic, and other problems |
| 2 | 1 | Widespread poverty throughout the world |
| 3 | 1 | Increase in the production, destructive capacity, and accessibility of all weapons of war |
| 4 | 1 | Uncontrolled urban spread |
| 5 | 1 | Generalized and growing malnutrition |

MENU    LIST ORDER

FIGURE 16

Prioritization is an important step for a very practical reason: highly preferred problems are of some central concern and should be addressed early in the process of building a structure. In this way, participants are relieved of the worry that their highly preferred issues might not get the attention that they deserve. At the same time, ideas which do not capture a high level of preliminary preference votes may actually be highly influential – meaning that ideas with few votes may turn out to be deeply buried (e.g., at the root of) the structure that the group constructs. Prioritization tells us only where to start during the structuring task.

MAPPING

**At this time, go to the MENU button (in Figure 3), call up the NAVIGATION tab, and select the MAPPING sub-program. On this new screen, select the option to "Begin a New Map."**

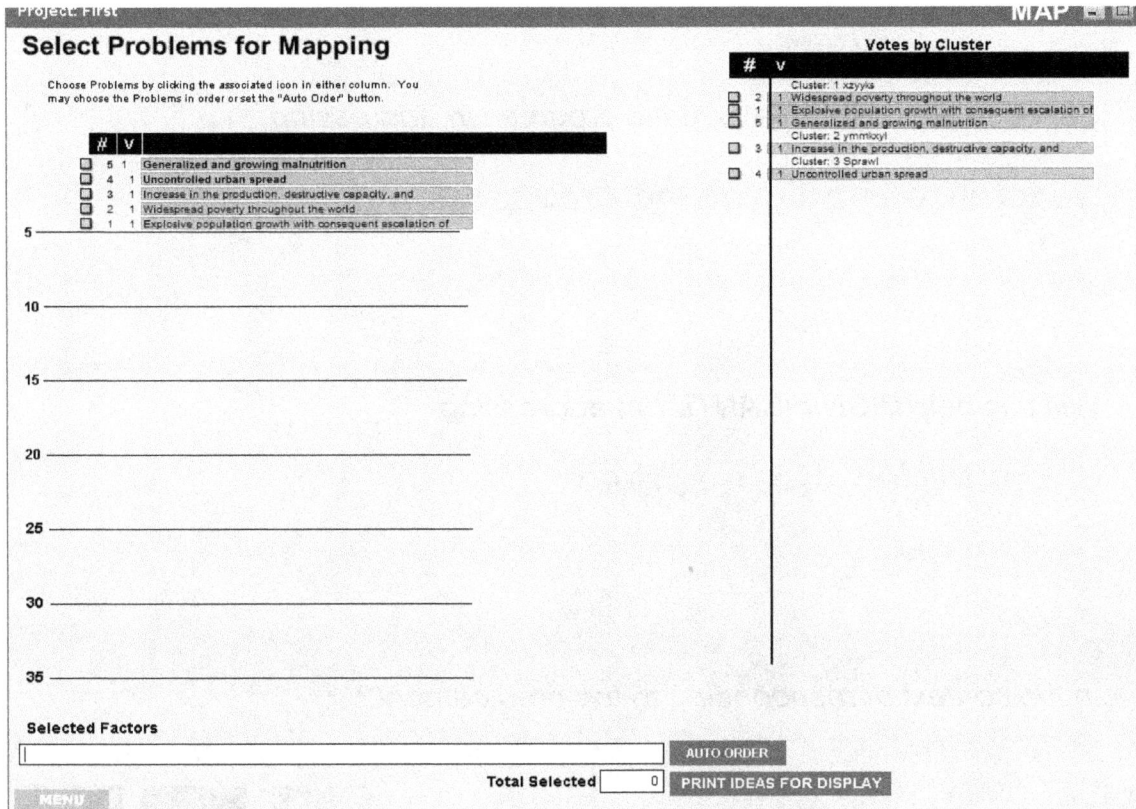

FIGURE 17

The resulting screen is titled "Select Problems for Mapping" (Figure 17). You may be asking why we need to select again after we have just collected preference votes. The screen is intended to support large group work. Where we have only five selected items to begin our map here; when we have a very large set of ideas all of which received different numbers of votes, this selection will be important in pulling the highest preference ideas into beginning of the structuring task.

**Your task here is to – ever so gently – put a check mark into each of the boxes next to the highly preferred problems that you will be entering into your preliminary individual map. Once you have selected the problems for mapping, a "Begin Mapping" button will appear in the lower right of the screen. Click on this button to launch the mapping sub-program that brings up the mapping screen.**

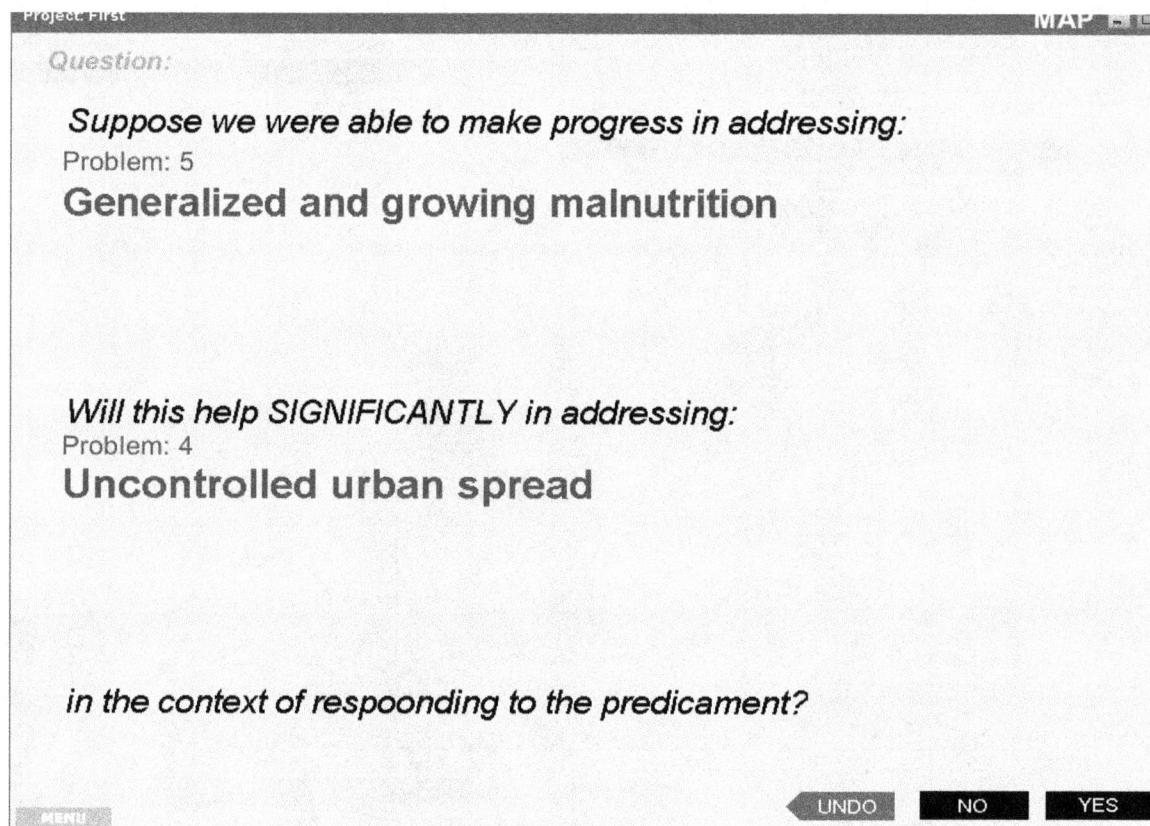

Project: First                                                                    MAP

Question:

Suppose we were able to make progress in addressing:
Problem: 5
**Generalized and growing malnutrition**

Will this help SIGNIFICANTLY in addressing:
Problem: 4
**Uncontrolled urban spread**

in the context of respoonding to the predicament?

MENU                                              UNDO    NO    YES

FIGURE 18.

The mapping screen is surprising to many who see if for a first time (Figure 18). It is surprising because it is a textual mapping tool and some individuals rather expect to see a graphic mapping tool. The program will automatically convert the textual map into a graphic map, and we will see this shortly. At this point we are prompted to make a logical assessment. **We are asked to use our body wisdom and our experience to determine if success in addressing problem #5 will SIGNIFICANTLY help us address problem #4 in terms of responding to the problem situation.** Sometimes a response comes quickly, and at other times responders feel themselves struggling with the question. **The question is asked in a specific order and it is highly important that the direction of the logical assessment is preserved.** Yes, we will ask the question in the reverse order (if rules of logic dictate that this should be done). The software is applying rules of "transitive" logic as it prompts us to make assessments, and the software is then recording our answers. That is all. The thinking and deciding is entirely our task. So ... what will it be? Think carefully. **There is an UNDO button on the screen, but it only allows you to undo one decision, so**

**take your time.** (Yes, there are ways for those skilled in the use of the CogniScope software to rapidly accommodate a group if a group should be reconsidering its decisions, but this advanced editing of decisions is beyond the scope of this introduction to the methodology and the software).

**When you have reached a decision, click the "NO" or the "YES" button and you will be prompted with another directional assessment with a pair of problems.** As you can imagine, in a group there may be a mix of no and yes votes. When this happens (as you shall see in Chapter 5), the dialogue manager needs to facilitate discussion of the assessment. If a supermajority (>66%) of the group sees the SIGNIFICANT influence in the direction specified by the question, and if none of the participants holds very strong objections to the group decision, the dialogue manager declares that the group has recognized a SIGNIFICANT influence and enters a YES vote. Anything short of this strong consensus results in a NO vote, and while the problem remains in the data set and is available to link into the structure at a later time, a decision has been reached and recorded with respect to where that specific component will not appear in the structure.

As you continue to work your way through the pair-wise comparisons that you are prompted to make, you will periodically be reminded that a new problem is being considered for the first time in this task. This helps groups remain attentive to a need for new distinctions. A prompt of this type is shown in Figure 19.

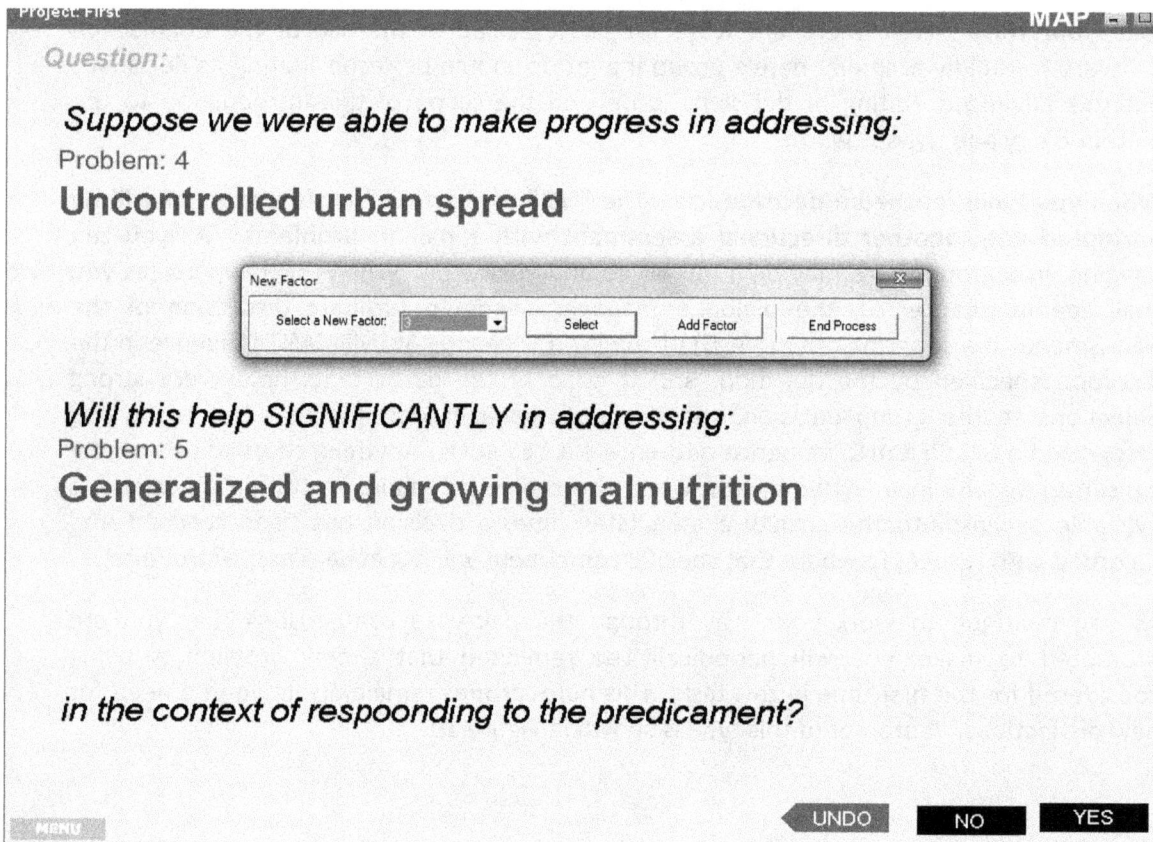

FIGURE 19.

**When prompted that a "new factor" (new problem) is being considered in the mapping task, you should choose to "Select" that new item.** For your current work, continue mapping until you are given the message: "Mapping process complete. All factors used." Then you are asked if you want to end process or add a new factor. **At this point, select "End Process."**

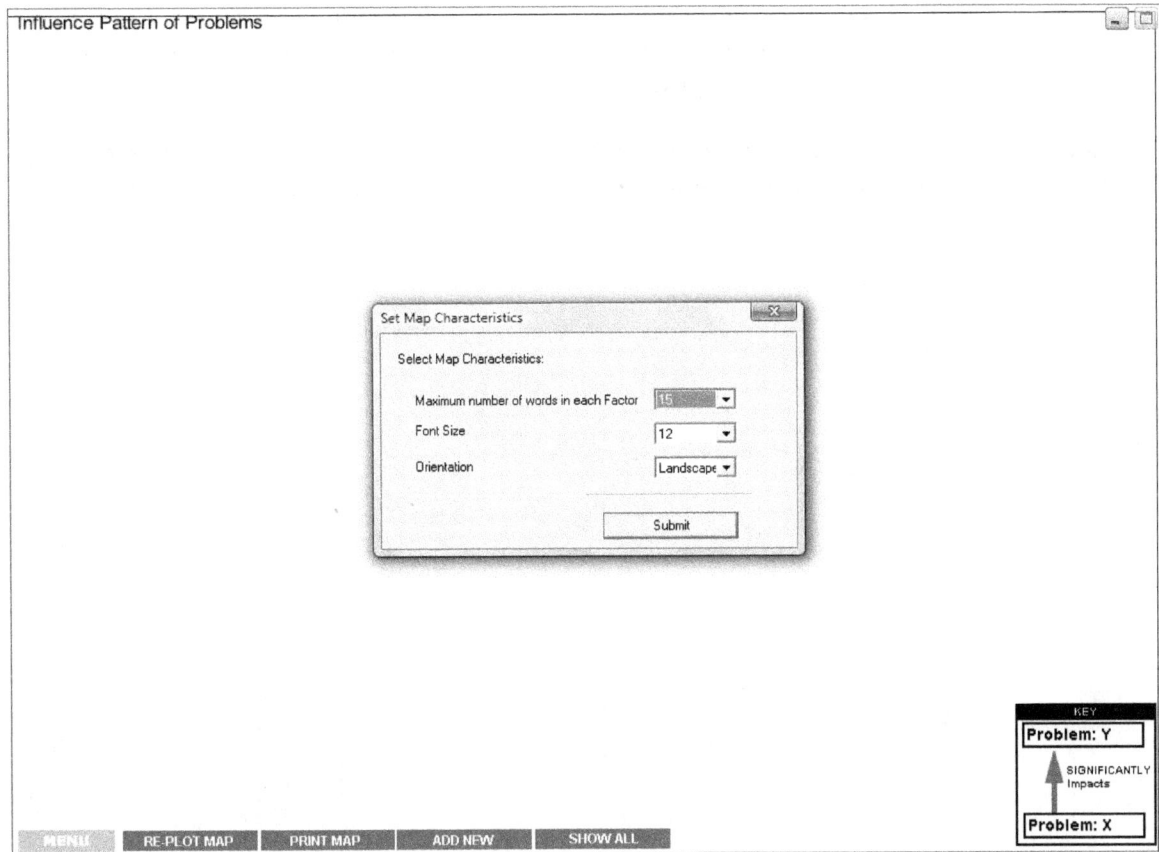

FIGURE 20

**For the task we are doing in this workbook, the default settings for generating a map are fine. Select "Submit." Your map will appear shortly.**

We are not going to show you the map here. You have already seen an example of the map that results from using interpretive structural modeling in Chapter 2. Your map will reflect your individual view. It may or may not have high correspondence with others in your class. What is sought is a map that will provide a basis of focused and open discussion with your fellow students.

When you see your map, you may edit it by "grabbing" and moving the text boxes in horizontal directions. The UTILITY tab below the map will provide you with options to edit the map. While this is an advanced function, it is worth mentioning here. From the new set of tabs that appear when the UTILITY function is selected, you can re-plot the map, print the map, or add new factors.

One final word.  An advanced utility function in the CogniScope software allows for rapid importing of data into clarification fields within the GENERATION sub-program, this function is well beyond the applications that we are introducing to you at this level.

**Conclude and save your work with the INDIVIDUAL2 file that you have created, open the REPORTS tab, and print out a set of reports to bring into the class to share and to discuss.**

# Chapter 5

## Using Structural Inquiry as a Group Process

This section of the sustainability class workbook takes on the challenge of applying structural inquiry as a group process. You have already been exposed to a mechanical understanding of how structured dialogic design is applied through the use of the CogniSystem software. You may also realize that you do not need this understanding to participate as a contributing expert in the structured dialogic design session. The background helps you understand the steps in the logical use of dialogue to explore complex situations – and it is essential if you are going to be the one who is using the software with a group. This chapter helps you understand some of the finer points of managing a complex dialogue as you work with the 49 problems from the global problematique.

The approach presented in this chapter has deep similarities with approaches used by professional design managers applying **Structured Dialogic Design**<sup>SM</sup>. Significant differences in "real world" practice and classroom exercises relate to:

- the framing of trigger questions (to evoke a list of critically important ideas),
- the recruitment of expert witnesses (to explain ideas with first person experiences),
- the overarching purpose of producing action plans so that the problematique can be resolved.

In classroom uses, the expected outcome is:

- an understanding of the complexity of global sustainability,
- an introduction to one means of addressing the complexity, and
- cultivating your will and confidence for addressing complex civic problems in your own lives.

### THE ESSENTIAL REQUIREMENT FOR DIALOGUE

As we have stated before, we believe that sustainability is not simply a matter of making changes. Rather it is more fundamentally a matter of making changes together. Acting together and making changes together – especially when the changes affect personal behaviors and lifestyles – require us to share a deep feeling that the changes are necessary and appropriate. This means that we need to make decisions together – and to make decisions together, we need to use dialogue.

You have been convened into a class project based upon your instructor's understanding of a need. This need may not have been apparent to you – and may still seem a bit mysterious. You are not alone.  Many individuals who are recruited into a dialogue for the purposes of building a consensus have different understandings of why the dialogue is important.

This diversity of perspective is found, for example, among the experts considering the problem. Typically, when technical experts are convened in a dialogue their first impulse is to see the dialogue as a vehicle for getting their individual perspective understood by others. Experts – as a consequence of their responsibility to express complex understandings – typically feel that they are called upon to teach more than to learn.  Without maligning experts, let us agree that this is a reasonable starting assumption.   The significant challenge for discussion among experts in different fields is to listen to and genuinely value the experts in other fields – including graduates from the schools of "hard knocks" and "street smarts."   A posture of gracious humility allows individuals to agree that they can learn from each other.  When such a communal feeling has been constructed, then authentic learning begins.

GROUP LEARNING BEGINS WITH A QUESTION THAT WE AGREE TO SHARE

 Let us assume first that we all agree that we can learn something new and important from each other – if not from each individual in our group then at least from the group as a whole.  Our first challenge is to focus our attention on a specific learning challenge.  In structured dialogic design, the focus is shaped by a "trigger question."  All information that is brought into the dialogue must be responsive to the trigger question.  Ideas which do not respond to the trigger question are extraneous – which is to say that they can distract the group and lead to a loss of group capacity. It is critically important that the trigger question be on target.

EXERCISE:  Consider the question *"What are the critical problems that we need to keep in mind as we seek to address the challenge of global sustainability?"*

As you consider this question, turn to Chapter 2 and examine **Table 1.  Continuous Critical Problems (CCPs): An Illustrative List.**   In this list you recognize labels for critical problems which were identified over forty years ago.  The labels, of course, do not tell the full story.  And in truth, only the author of the labels could tell us exactly what stands behind the label.  The author is no longer with us.  We are challenged to investigate what we feel these labels may mean in our now modern situation.

Do you accept this as a reasonable question?  If you do not accept this question, what might you propose to make the question more acceptable to you AND the rest of the group?

Accepting the question for joint exploration is part of the group decision making process.  In our classroom situation devoted to examining global sustainability, we have adopted Hasan's question.  If we were convened to engage a different complex situation, the convener would introduce, explain, and possibly revise the trigger question.  As a matter of historical reference,

Hasan Ozbekhan was asked to answer this question when he was charged to design a prospectus for the Club of Rome. We feel that the question is succinct and timeless. Do you agree?

EXPERTS CONTRIBUTE IDEAS INTO THE SHARED UNDERSTANDING

[The "experts" for this class are you the students. In Chapter 2, you were assigned the CCPs for which you are the designated experts in the ongoing discussion]

GATHERING AND CLARIFYING IDEAS

[In this class, the ideas are already gathered. They are the 49 CCPs. The following text, prior to discussing the classification. is presented for completeness of understanding]

When groups are learning from their members, in is best if the group can be organized in a circle with all members of the group looking toward all other members of the group. We cannot say if this ideal situation exists in your classroom. If you are using this approach online, we recognize that you will be talking to each other "through the Internet" and you will be missing out on many cues of facial expression and body language. You will be missing further cues if you cannot share a conference call. Dialogue is intended to make use of multiple dimensions of expression, and where dialogue involves sharing body wisdom, multi-channel communication is really very important.

If you are not meeting in a face-to-face classroom, then you will be relying upon your instructor to guide you to a wiki (or a wiki-like learning platform) where you can post questions, clarifications, questions about clarifications, and – well – clarifications of those questions. We will not discuss the online wiki methodology in this chapter, however, we will provide a brief appendix as an aid to instructors at the end of this workbook.

When you are assembled in your best possible approximation of a circle, ideas for the group's consideration should be gathered and displayed on the walls of the room in such a fashion that all can read the ideas. When all the ideas are gathered, they are also presented to each participant as a compiled list. Structured dialogic design gathers up ideas in a round-robin fashion so that each group member contributes one idea in turn until all ideas have been harvested. As members of the class identify an idea (a "problem") that they have studied, that idea should be visible to all others as a display on the wall or within the wiki. Once all idea labels have been posted, a cycle of clarification begins.

[In our classroom project the ideas on the wall and the compiled list should be prepared for display ahead of time]

Each labeled idea is revisited in turn, and the "author" of the label for the specific problem is asked to provide a brief statement of the nature of the problem. The author's clarification is captured and displayed too. On the wiki, each student posts their own statement and in a face-to-face setting, some member of the class sits with the CogniSystem software and enters clarifications as authors contribute them.  With 49 problems to consider, this can (and genuinely should) take a good span of class time.   In an effort to use precious classroom contact time most judiciously, we recommend the use of an online wiki as a means of gathering and displaying the set of 49 problems before class – with the opportunity to project the wiki in class and then combine an oral clarification with the posted clarification.

The goal in this workbook is to allow student "experts" to be directly responsible for a small set of problems and yet also to assure that all students will be "aware" of the full set of problems. Students who post labels and clarifications for ideas to the wiki will have demonstrated their individual responsibility for their set of problems.  Students who pose questions for clarification for specific problem posts will have demonstrated their group awareness of problems.  To manage the back-and-forth flow of postings, each student might be asked to post their five (or three) problems, and to assist clarification of three (or five) other postings.  The wiki will track the names of all authors and all students participating in group clarification.

In our experience, while some students may be asked for a clarification by many other students, a single clarification response is generally sufficient to address many questions.

In a classroom setting where clarifications are provided verbally in the ideal face-to-face fashion, and where an individual serves as the "recorder" to capture the essence of the clarifications, a printed record of the clarification should be rapidly produced and distributed to all participants for their future tasks.

Clarification is continued only as long as group members feel that clarification is needed.

NOTE:   During clarification, dialogue managers need to enforce one simple rule: **"no value judgment for or against any of the ideas which are presented during the clarification phase."** The goal of the clarification task is to UNDERSTAND a perspective, even if a perspective may be UNPOPULAR.  It is fair to ask for examples of problems to better understand problems.  It is fair to offer examples of problems to check-in and see if meanings are being shared accurately.  It is not fair to tell an "author" who owns a problem what someone else feels that that problem REALLY means or SHOULD mean.  If someone feels that the essence of problem is not being expressed, then a new idea may actually have been discovered from within the group.

NOTE 2:  Recognize that "meaning unfolds" and that new dimensions of meaning will emerge as ideas are considered in different ways during a complex dialogue.  Do not try to capture all meaning at one glance.  Grab and present the essence of the meaning.  Even though a student may spend two hours "researching" a complex problem, the goal is to capture and share the essence of that problem as efficiently as a student possibly can do this.  Too many words will exhaust the audience.  It may be that younger audiences who are currently building capacities to communicate through text tweets will demonstrate profound skill in telegraphing meaning among groups.

---

**EXERCISE:  Reflect on your participation in a clarification session.**

Structured dialogic design immerses you in complexity.   You are expected to become "overloaded" with information.  Your challenge is to keep your assigned problems clearly in mind and to explore the vastness of the collection of ideas brought to the table by the rest of the group.  As each group member works to "protect" the understanding of their own problems, some ideas will begin to collide with each other.  You can expect your classmates to ask if your problem is different from a problem that they have researched – and you may or may not have answers for all of the questions that you get.

In response to the give and take of clarification, do you feel that you understand your problems better?  Do you feel that you have learned how to express your problems more effectively?

---

## CLUSTERING IDEAS INTO SETS WITH SHARED ATTRIBUTES

Clustering is very, very difficult to do when group members are not engaged face-to-face.  The method for clustering as a result of individual assessments has been described in Chapter 4.  With a group, assessments of similarity and difference results in some head shaking and head nodding even before concrete statements to substantiate choices emerge.  Different individuals see different attributes and weigh the different attributes in different ways.  Some fits are loose and strained while others are tight and simple.  In face-to-face work, groups understand the struggles that they are undertaking and find ways to help each other.  Virtual groups lack this dynamic with currently available technical support systems.  For this reason, we recommend that a subset of group members who can interact directly agree to construct a "proxy" clustering.  Their work is then shared with the larger group who respond with specific suggestions for revisions as they feel may be needed.

We have additionally considered holding 20% of the problem set (e.g., 9 or 10 of the 49 continuous critical problems) to be "added in" to the proxy cluster with input from the full group.  This part of structured dialogic design contributes richly to deep understanding of the issues the group is working with, so accommodating the mental activity of clustering in the best fashion that one can is important.

Once the group has agreed upon the affinity cluster pattern, each cluster is named.  As the label is place on the cluster, the group captures a "higher order" awareness of the problems they are seeking to understand.  The higher order suggests "dimensions" of the problem situation … where a dimension can reflect a perspective or a "way" of looking at subsets of the problem situation.

---

**EXERCISE:  Reflect on the clusters accepted by the group.**

In the world of taxonomies, there are "groupers" and there are "splitters."  Do you feel that the ideas (problems) were spread out too thinly (split) or too sparsely (grouped)?  Do you feel that the clusters provide a new and useful means of understanding the problems that you are investigating?

Do you feel yourself being drawn toward specific clusters with a sense that some clusters might address urgency or importance more strongly than other clusters?

---

VOTE ON PREFERENCE BASED ON INDIVIDUAL SUBJECTIVE ASSESSMENTS

In a group setting, if no learning has occurred, then each group member votes for importance of the ideas which they have contributed.  Thus, votes would be spread across all ideas evenly.  Where individuals are invited to initially contribute their top five ideas for group consideration, preference voting with five votes could result in situations where no preferences emerge.  This has never been observed to be the case.  Evidence of learning has always been reported.  Preference votes converge and preferences emerge.  In complex dialogues, convergence also has always been observed to be incomplete.  With options to vote for five highly preferred ideas to be included in a preliminary structure of a complex situation, never have only five ideas been identified as being centrally important by a group.  The situation is more typically consistent with the 80/20 rule, which conjectures that 20 percent of the ideas will account for 80 percent of the votes.

Preference voting has a cathartic effect.  While not attacking less popular ideas, the group focuses first on its strongest basis for agreement.  Because the preference votes are individually cast without group input (beyond accumulating evidence of where others may already have cast their votes), this task is not an authentic group decision making task.  It remains an aggregation of individual decisions.  As such, preference voting results is a premature reflection of shared understanding.  The catharsis that accompanies voting for one's preferences unfortunately also conveys a false sense of "completion."

**EXERCISE:  Reflect on preference voting.**

Consider the distribution of your preference votes in the context of the group results.  Did your votes accumulate to any of the top 5% of the group's aggregated preferences?  Do you feel that your preferences are reasonably represented by the group data?  Do you sense that the group work is "done?"

Consider the ideas that you originally brought to the table.  Have any of them risen to be recognized as highly preferred through the aggregated votes of the group?  Do you feel that some of the non-preferred ideas should have received more votes?  Though the voting may have been a "democratic" expression, do you feel that the group genuinely understands its most strongly shared priorities?

STRUCTURING PREFERRED IDEAS

Chapter 4 presented the method for structuring preferred ideas using the Interpretive Structural Modeling (ISM) routine of the CogniScope software.  Prior to running the ISM, handouts of the ideas generated, their clarifications, and the tally of preference votes are distributed to the members of the class.  The structuring session is guided by a generic question which frames comparisons for each pair of ideas.  In response to the question a vote is taken.  The vote is a "public vote" – meaning that voters are expected to share their reasoning with each other.  Secret or anonymous votes would be counter- productive.

It is acknowledged, however, that with some subject matter, public voting and open discussion of politically sensitive ideas may put some participants in a structured dialogue at personal risk after the dialogue has closed and participants return to prior work environments.  While individual students do not put themselves at risk in naming issues which are matters of historical reference, their stated contemporary beliefs about influences among ideas may reveal personal values that – in repressive political environments – could expose them to retributions subsequently.  The use of structured dialogic in open democratic discussions presupposes that cultural conditions tolerate (or authentically espouse) democratic principles across race, creed and gender.  International participants who are invited into online classes using approaches provided in this text are encouraged to be sensitive to special cultural needs.  We recommend distinctions between "participant" roles and "observer" roles when active participation in democratic dialogue might be constrained due to political sensitivities.  This said, we do encourage adoption of authentic democratic practice as the most robust and promising means of resolving truly complex problems involving human systems.

In response to the generic question asking participants to assess the directional influence that resolving one specified problem would have toward resolving another specified problem, votes are first taken by a show of hands.  Online, this can be captured as a stream of chat box comments of YES or NO.  Early and emphatic responders are recognized as participants with strong views.  If the voting is split, the strong responder from the minority point of view (for example, an early NO in the context of many YES votes) is called upon to present his or her perspective on the influence relationship. A strong view from the majority perspective is then taken, and samples can be taken from both sides iteratively until a range of differing views have been voiced. Votes are taken again, and if the group is still largely divided in its view, more samples can be collected. After a second round of voting, if less than 67% (a supermajority) of the group agrees that a SIGNIFICANT influence exists, the conclusion is recorded that a significant influence was not recognized by the group as a whole.

At the conclusion of pair wise comparison, the structure of the group decisions is presented to the group for its evaluation. The structure is generated automatically as presented in Chapter 4.

NOTE: Voting must proceed with care and respect. It cannot be rushed, and explicit efforts should be taken at the start to illustrate that the dialogue manager will not allow the vote to be rushed. The votes also should be recorded through an open counting process. The show of hands in public voting assures that the group will monitor the integrity of the vote tallying. Unlike preference voting (which was an aggregation of individual votes), this voting is group voting and the record of the group vote is central to the group's sense of legitimacy.

---

**EXERCISE: Reflect on the group structure.**

The structure produced by the group using Interpretive Structural Modeling is a learning artifact. This is to say that the structure summarizes what the group has learned about the influence the problems it has considered have on the global problem of sustainably.

Consider your feelings as you first saw the structure. Were your eyes racing across the structure of signific ant parts of the landscape? Did you sense importance? Did you sense mystery? Did you feel that the structure was an authentic expression from the group?

---

AUGMENTING THE STRUCTURE OF THE GROUP UNDERSTANDING

The Interpretive Structural Model is a living model. It is not complete simply because it was generated automatically. The group will "read" the model (with deep drivers of the influence among problems in the problematique propagating from the roots and moving upward through the branches). The group will identify bottlenecks in the structure. And the group will recognize highly influenced yet weakly influential problems at the top of the structure.

As the group reflects on its strong basis for agreement, the group should be invited to also consider those ideas which are not yet in the map. While all of the ideas can be woven into a map if a group elects to do this, groups generally choose to scan the "residual" ideas in search of idea which might be more deeply positioned in the tree structure that they have created. This is to say that the group becomes tuned to seeking out and recognizing highly influential ideas which, were they resolved, might exert strong "leverage" across the problematique.

This part of the group process can be managed easily if a wall display is created showing the tree with its component ideas, and inviting the group to test specific additional ideas for their fit beneath the deepest driver in the structure. In this way, Individuals and the group share a respect for systemic problems and a desire to address them.

CONSIDERING THE AUGMENTED STRUCTURE IN RELATION TO INITIAL PREFERENCE VOTES

We find it is valuable to remind a group of what it has learned. Even with the discovery of a structure that no individual could singly have predicted as a group outcome, a group may miss the subtle yet profound realization that its initial preference votes would – if embraced – have burdened the group with a false sense of priorities. This statement is made with a strong basis of empirical knowledge. Preliminary preference votes in group decision making were compared directly with priorities that arose from Interpretive Structural Modeling in a complex situation. The results confirm what you have found in this project: the influential factors are far different than those judged most important in individual voting. The misdirection has been called the Erroneous Priorities Effect (see Dye and Conaway, 1999 wherein popular votes consistently fail to correspond with deep drivers in an influence map (Figure 1; reproduced with permission of authors[1]).

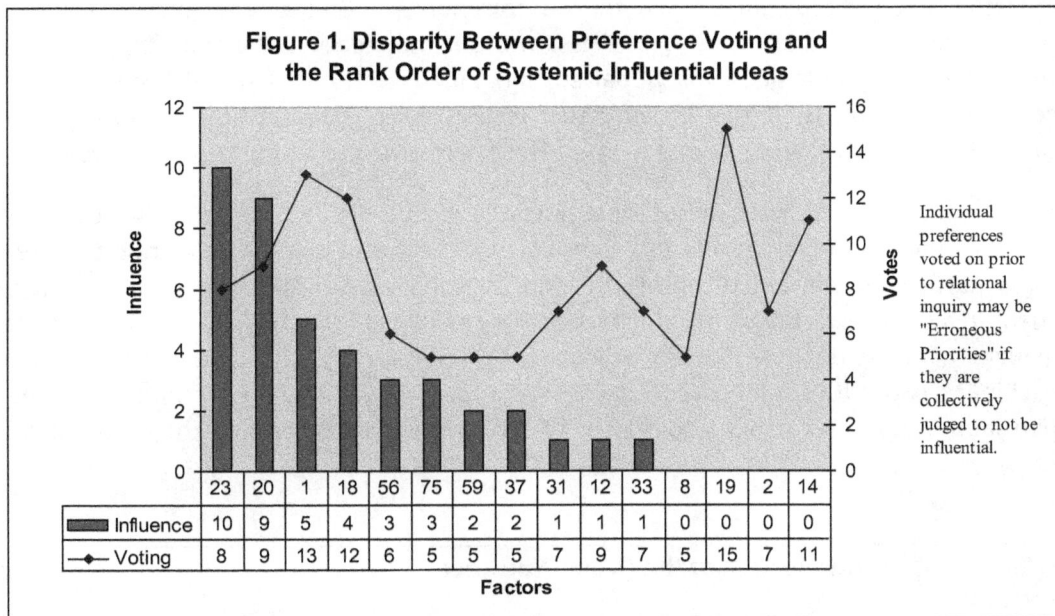

Figure 1. Disparity Between Preference Voting and the Rank Order of Systemic Influential Ideas

| Factors | 23 | 20 | 1 | 18 | 56 | 75 | 59 | 37 | 31 | 12 | 33 | 8 | 19 | 2 | 14 |
|---|---|---|---|---|---|---|---|---|---|---|---|---|---|---|---|
| Influence | 10 | 9 | 5 | 4 | 3 | 3 | 2 | 2 | 1 | 1 | 1 | 0 | 0 | 0 | 0 |
| Voting | 8 | 9 | 13 | 12 | 6 | 5 | 5 | 5 | 7 | 9 | 7 | 5 | 15 | 7 | 11 |

Individual preferences voted on prior to relational inquiry may be "Erroneous Priorities" if they are collectively judged to not be influential.

[1] **Dye, K.M. and Conaway, D.S.** (1999), 'Lessons Learned from Five Years of Application of the CogniScope™ Approach to the Food and Drug Administration', CWA Report (Paoli, Pennsylvania; Interactive Management Consultants).

---

**EXERCISE:  Define Your Group Learning.**

Preference voting gives rise to erroneous priorities.  Reflect upon the ideas (problems) which your group identified as being of highest priority based upon aggregation of individual votes.  Compare the ranking of the popularity of ideas (the number of preference votes an idea got) with ideas which had the greatest systemic influence (count the number of ideas which are connected above a specific idea in your map).

Human beings, as nomadic hunters, can be thought to have an innate sense of immediacy and bottleneck – proximity and entrapment.  There is little evidence for an innate sense of "leverage" in a complex system.  Do you feel that this class exercise has enriched your sense of leverage in the context of global problems?

---

## NARRATIVE MANAGEMENT WITH STRUCTURED DIALOGIC DESIGN

Participants in group design co-create artifacts which lead to new substance in the world.  However, the translation of design to enactment is not guaranteed.  To manage a transition, a new narrative needs to be introduced to the community.  One of the explicit goals of the application of structured dialogic design to classrooms is the creation of a story that can lead to a resolution of the problematique for global sustainability.   There is no ONE narrative that is ordained to be the Holy Grail for all groups to discover.  Each group must discover and enact its own narrative.  To do this, groups need to voice their narrative as that narrative is discovered.

Expressing a narrative is as much about theater as it is about science.  The script is provided in the structure that the group creates and the learning that the group has shared during this creation.  Each class participant can tell this story in individualized ways.  This is encouraged, for each participant in a complex dialogue reports back to their individual constituencies and must speak in the language of those people.  The act of voicing complementary narratives based on the same script achieves two critical objectives for global sustainability: first, it puts into voice the new narrative; and second, it fosters a tolerance of multiple voices describing the true narrative in distinct_ways.

---

**EXERCISE:  Present Your Version of the Group Narrative.**

Interpretive Structural Model (ISM) maps can be read in many ways.  Convention, however, has the reading start with the deepest driver (the bottom most) idea and then reporting influence as it flows upward through the "tree."  Some audiences might be willing and able to understand only a

---

portion of the tree. This presents a challenge in communicating a truthful narrative in its most parsimonious yet faithful form.

Define the audience for whom you would feel a need to share the class' narrative. Present your version of that narrative. Can you envision ways to enrich your narrative with symbolism or with art?

## PROCESS REFLECTION

You have now completed an introductory course on the application of structured dialogic design to the problematique for global sustainability. In the process of experiencing this approach, you have learned about 1) continuous critical problems in the world, 2) cultivating a group understanding of its situation in the world, and 3) applying a formal methodology to engage complex problems involving human systems.

**FINAL EXERCISE: How Will You Apply What You Have Learned?**

Interpretative Structural Modeling is a powerful approach. It is also underutilized as a social problem solving methodology. Based on your experiences beyond the classroom, what large group design and problem solving approaches have you experienced? How do they compare?

Hint: The answer is NOT none.

# Chapter 6

## Illustrations of Continuous Critical Problems

In this section of the sustainability workbook, we offer some perspectives on the 49 continuous critical problems (CCP) which have been introduced earlier in Chapter 1 and 2. Our goal here is to illustrate how (on the date that this chapter was constructed) current events reflect examples of each of the CCPs. Naturally, when you consider the CCPs that your group has been assigned by your teacher, you will look for examples in the most current news that you can find on the Internet. For this reason, our examples will illustrate how we searched for recent examples. You are likely to have better information at your finger tips as you take on the task of understanding CCPs, so your examples are likely to be superior examples.

The overarching goal for you and your team is to become the local expert on current examples of CCPs that have been assigned to you so that you can explain the CCPs to your fellow classmates. Through your well researched explanations, you will be able to help others understand what you feel your CCPs "mean." It is all about meaning, so if you manage your research well, you will have rich meanings to share with others. You cannot accept even the sites which we provide as examples as being unequivocal "truths" – we are not trying to deceive anyone, however, on occasion we can offer opinion to illustrate beliefs yet without resources to establish the validity of such belief. Beliefs themselves are, of course, political realities. You must weigh the evidence you find and discover your sense of truth. And there is a value to this task which transcends this course: performing online searches for meanings will prove invaluable to many of the problems that you face in your professional and personal lives in the future.

---

**EXERCISE**

Capture a big picture first and then "boil down" your understanding of the continuous critical problems into a concise statement for sharing with your class. To do this, compose a "short report" from your research – two or three pages at the most – and then reduce the report to a single paragraph (with a few well selected websites) which will reflect the "essence" of your understanding. If your class chooses to collect and grade full reports, you will be assured that comparable effort has been invested in the task by all class participants (which, of course, is not a significant concern in mature classes). If concise expression is not your strength, you might choose to supplement your concise statement with your short report to assist you in earning as much academic credit as you are seeking from the class.

*Note: We believe that we are all partners in understanding our world. If you choose, you may submit your summary of specific problems to us and, if we include your summary in a future version of this book, we will recognize you as a "contributing author." Please do include complete return addresses with submissions.*

## 49 Continuous Critical Problems with Selected Websites

[Note: *these are not complete or preferred answers, but illustrate approaches to understanding the problems*]

### 1. Explosive population growth with consequent escalation of social, economic and other problems

Overview of Population Growth Rates
http://geography.about.com/od/populationgeography/a/populationgrow.htm If Afghanistan's growth rate remained the same (which is very unlikely and the country's projected growth rate for 2025 is a mere 2.3%), then the population of 30 million ... Population growth rate http://www.eoearth.org/article/Population_growth_rate The population growth rate measures how populations change in size over time. The units of population growth rate are individuals per time

Human Population Growth
http://users.rcn.com/jkimball.ma.ultranet/BiologyPages/P/Populations.html The Rate of Natural Increase Birth rate (b) − death rate (d) = rate of natural increase (r)

Population Profile of the United States http://www.census.gov/population/www/pop-profile/natproj.html The U.S. population growth rate is slowing, despite large increases in the number of persons in the population

Principles of Population Growth http://waynesword.palomar.edu/lmexer9.htm It has been clearly documented that when a nation's standard of living and gross national product (GNP) increases, its population growth rate actually declines

## 2. Widespread poverty throughout the world

Causes of Poverty  http://www.gdrc.org/icm/poverty-causes.htm  Poverty is explained by general ... entrenched sources of poverty throughout the world ...

The Myth of Widespread American Poverty
http://www.heritage.org/Research/Reports/1998/09/The-Myth-of-Widespread-American-Poverty  The average "poor" American has one-third more living space than the average Japanese does and four times as much living space as the average Russian
Poverty Around The World — Global Issues
http://www.globalissues.org/article/4/poverty-around-the-world ... some of the causes of poverty around the world ... previous wars

BBC NEWS | World poverty 'more widespread'
http://news.bbc.co.uk/2/hi/business/7583719.stm
The World Bank has warned that world poverty is much  'more widespread'

LankaWeb – A Brief Inquiry  http://www.lankaweb.com/news/items/2010/12/29/a-brief-inquiry-into-the-continued-widespread-poverty-tragedy/  injustice and mass poverty throughout the world

Research Paper  http://www.essaytown.com/paper/poverty-childrens-education-widespread-poverty-severely-affects-education-28177  100 million children throughout the world

## 3.  Increase in the Production, destructive capacity, and accessibility of all weapons of war

Weapons of the cold war  http://www.is.wayne.edu/mnissani/pagepub/CH4.html ... this century, dramatic increases in the technological sophistication and effectiveness of conventional weapons have taken place

Rape is increasing as a weapon of war  http://www.english.rfi.fr/africa/20101020-empowering-women-can-help-prevent-conflict-says-un-report  Conflict and war is less about traditional battle lines and national borders, but "more about combatants struggling for control within a single country and employing any means to break the will of civilians"

Chemical Weapons  http://www.globalsecurity.org/wmd/intro/cw.htm  forswearing only the first use of the weapons and reserved the right to retaliate in kind if chemical weapons were used against them

World War Two http://www.bayonetstrength.150m.com/Weapons/infantry_weapons_of_world_war_tw.htm  saw a massive increase in the variety of weapons used to equip the average infantry battalion

Sales of Weapons http://video.search.yahoo.com/search/video?p=increase+in+weapons+of+war  A Surge in U.S. Weapon Sales to the World

## 4. Uncontrolled Urban Spread

Urban sprawl Facts  http://www.encyclopedia.com/topic/Urban_sprawl.aspx  Urban sprawl may be defined as the low-density, haphazard housing development that spreads out around modern towns and cities

Urban Sprawl http://www.georgiaencyclopedia.org/nge/Article.jsp?id=h-763  The environmental impacts of urban sprawl in Georgia are among the most significant and widespread in the nation

Sierra Club Stopping Sprawl  http://www.sierraclub.org/sprawl/  The Challenge to the Sprawl Campaign works to fight poorly planned runaway development and promotes smart growth communities

Urban sprawl  http://www.absoluteastronomy.com/topics/Urban_sprawl also known as suburban sprawl, is a multifaceted concept, which includes the spreading outwards of a city and its suburbs

The Environmental Literacy Council - Urban Sprawl
Urban Sprawl. Within the United States, in the decades following World War II, rising levels of prosperity, the widespread availability of affordable housing and ...

## 5. Generalized and growing malnutrition

Effects of malnutrition on personality http://findarticles.com/p/articles/mi_m0887/is_n3_v10/ai_10600724/ six studies of the effects of generalized malnutrition / semi-starvation

Socioeconomic inequality http://www.who.int/bulletin/volumes/86/4/07-044800/en/index.html different patterns found for the distribution of malnutrition across socioeconomic groups

Protein & Malnutrition http://www.livestrong.com/article/303024-protein-malnutrition-disease/ can cause changes in skin pigment, decreased muscle mass, diarrhea, failure to grow, fatigue, changes in hair color or texture, irritability, rashes, shock and generalized swelling, or edema

Disproportional burden http://www.ncbi.nlm.nih.gov/pmc/articles/PMC2245943/ related to poverty, maternal education, health care and family planning and regional characteristics

Facts and Statistics http://www.worldhunger.org/articles/Learn/world%20hunger%20facts%202002.htm cThe second type of malnutrition, also very important, is micronutrient (vitamin and mineral) deficiency

## 6. Persistence of widespread illiteracy

Widespread illiteracy http://www.nytimes.com/1986/07/22/science/education-widespread-illiteracy-burdens-the-nation.html as many as one-third of the adults in the country are essentially illiterate

Literacy Crisis http://www.articlesbase.com/k-12-education-articles/widespread-illiteracy-ending-our-literacy-crisis-1235825.html problems of functional illiteracy and the widespread extent of illiteracy

Widespread illiteracy http://www.independent.co.uk/opinion/letters/letters-widespread-illiteracy-891442.html even labouring work requires the filling in of forms and reading instructions

Newspaper Decline http://www.inteldaily.com/news/173/ARTICLE/12166/2009-10-09.html One reason for the decline of newspaper circulation is that 42M Americans are illiterate and roughly 50M more are semi-literate

Illiteracy Hurts Everyone http://searchwarp.com/swa551130-Widespread-English-Illiteracy-Hurts-Everyone-Seven-Vital-Questions.htm it is in the short-term best interests of political and educational authorities to downplay the seriousness of the English literacy problem

## 7. Expanding mechanization and bureaucratization of almost all human activity

Human Behavior in Organizations http://ezinearticles.com/?The-Human-Behavior-in-Organizations&id=2536182 Bureaucracy is a form of human organization that is based on rationality

Reinventing NASA http://www.amazon.com/Reinventing-NASA-Spaceflight-Bureaucracy-Politics/dp/0275970027 internal values, policy choices, and relations with other political players are all driven by its overriding goal

Human Nature http://www.nature.com/news/2009/090211/full/457780a.html People's mindsets are neither fixed by evolution nor infinitely malleable by culture

Bureaucracy of India http://www.scribd.com/doc/23540639/VALUE-SYSTEM-IN-INDIAN-BUREAUCRACY Contrarious values – inherent values and survival-oriented values clash

Wikipedia http://en.wikipedia.org/wiki/Bureaucracy Bureaucracy is the combined organizational structure, procedures, protocols, and set of regulations in place to manage activity, usually in large organizations ... as opposed to adhocracy

Thinking About Bureaucracy http://www.cobbers.org.au/bureaucracy.html the more remotely bureaucracy controls an area of a human activity, the less sensitive it is to local needs or particular cases

## 8. Growing inequalities in the distribution of wealth throughout the world

Wikipedia http://en.wikipedia.org/wiki/Wealth_inequality Economic inequality comprises all disparities in the distribution of economic assets and income.

Wealth inequality is vast and growing

Wealth inequality is vast and growing Sylvia A. Allegretto August 22, 2006 See Snapshots Archive. This week's Snapshot previews data to be presented as part of the ...

Wealth Inequality
http://www.telegraph.co.uk/finance/financetopics/davos/8283310/Davos-WEF-2011-Wealth-inequality-is-the-most-serious-challenge-for-the-world.html the most serious challenge facing the world in the years ahead

New York Times
http://www.nytimes.com/2010/10/17/business/17view.html?src=busln?_r=1 The share of total income going to the top 1 percent of earners, which stood at 8.9 percent in 1976, rose to 23.5 percent by 2007

Denver Post http://www.denverpost.com/nationworld/ci_16343819%3fsource=rss
Growing wealth gap hangs over tax-cut debate

The Huffington Post http://www.huffingtonpost.com/ray-brescia/reducing-income-inequalit_b_764477.html Reducing Income Inequality: The Next Phase of True Financial Reform

Wealth and Income Inequality
http://www.multinationalmonitor.org/mm2003/03may/may03interviewswolff.html
The Growing Gap in the United States

Inequality Is Widening Worldwide
http://www.nytimes.com/2006/12/06/business/worldbusiness/06wealth.html?_r=1
Income inequality is near record levels in many countries, and the world's wealth has become even more narrowly concentrated than income.

## 9. Insufficient and irrationally organized medical care

Wikipedia http://en.wikipedia.org/wiki/Healthcare_inequality Healthcare inequality refers to the disparities in the access to adequate healthcare between different gender, race, and socioeconomic groups

Inequalities in Health Care
http://www.willwilkinson.net/flybottle/2009/10/20/inequalities-in-health-care/ the only hope of eliminating the inequality is forbidding access to treatments that that cannot be provided to all under the universal health insurance system

Socio-economic inequality http://www.sociology.org/content/vol8.1/deogaonkar.html socially under-privileged are unable to access the healthcare due to geographical, social, economic or gender related distances

Health care quality http://www.stanford.edu/~pista/jpw.pdf self-insurance (saving) may be the only available o ption

Costs http://www.cliffsnotes.com/study_guide/Health-Care-Costs-and-Inequality.topicArticleId-26957,articleId-26938.html newer, more expensive drugs, particularly newer antibiotics and drug treatments for AIDS patients, also contribute substantially to rising costs

## 10. Hardening discrimination against minorities

Discrimination against Minorities http://www.ohchr.org/EN/AboutUs/Pages/DiscriminationAgainstMinorities.aspx Minority issues have been on the agenda of the United Nations for more than 60 years

Racial Minorities http://www.lotsofessays.com/viewpaper/1683883.html Discrimination in lending is a part of the larger problem of housing discrimination

Human Rights Watch http://www.hrw.org/en/news/2010/12/15/iran-discrimination-and-violence-against-sexual-minorities Abolishing Iran's discriminatory laws and policies is critical to ensuring protection of its vulnerable sexual minorities

Economic discrimination http://en.wikipedia.org/wiki/Economic_discrimination economic factors can include job availability, wages, the prices and/or availability of goods and services …

Minorities Discriminate Against Minorities http://www.ethnoconnect.com/html/articles_28.html Prejudice is about fear, fear of the unknown and fear of others who are different from ourselves

## 11. Hardening prejudices against different cultures

Prejudice between cultures http://psychlotron.org.uk/resources/social/AS_AQB_social_prejudiceculture.pdf All cultures seem to make a fundamental distinction between 'us' and 'them' and it appears universal that they favour 'us' over 'them' for many purposes

Prejudice http://www.newworldencyclopedia.org/entry/Prejudice   Colonialism was based, in part, on a lack of tolerance of different cultures

Anti-Prejudice http://www.themoreyouknow.com/Diversity_Anti_Prejudice/ Children recognize the differences between people at an early age. They begin to distinguish different skin colors and facial features at six months and start to understand their own individuality

Prejudice, Intolerance, and Nationalism http://www.islam4all.com/race_prejudice,_intolerance,_and_nationalism.htm intolerance is a local and temporal culture-trait like any other

Antibias Curriculum and Instruction  http://www.ericdigests.org/1997-2/antibias.htm celebrate--rather than denigrate--the diversity in American culture and language usage

## 12.  Affluence and its unknown consequences

Affluence, a Curse or a Blessing?
http://novice101.wordpress.com/2009/04/04/affluence-a-curse-or-a-blessing/
Affluence has also created a huge demand ...

Peak Oil Consequences  http://www.peak-oil-news.info/global/consequences/
Governments will need to invest at least $20B into the energy infrastructure over the next 25 years to meet the growing worldwide demand for electronic technologies and gadgets

Personal Bankruptcy http://www.startingovertoronto.com/articles/v08.php Insolvency practitioners are seeing an increase in problem debtors whose financial difficulties arise from poor spending habits, particularly compulsive spending and lack of financial discipline.

Satya Center http://www.satyacenter.com/news-global_visionaries-challenge-of-affluence Our historically rapid rise to great affluence poses a new, and as yet unmet, moral challenge to our culture

Barbaric Affluence http://www.michael-oakeshott-association.com/pdfs/conf06_mosley.pdf Thus was generated a new art, not of ruling, but of knowing what offer will collect most votes, and of making it in such a manner that it appears to come from 'the people'

## 13. Anachronistic and irrelevant education

School is boring and irrelevant
http://www.telegraph.co.uk/education/secondaryeducation/4297452/School-is-boring-and-irrelevant-say-teenagers.html The corrosive divide between academic and vocational learning, reflects a social attitude which views 'know-how' as inferior to 'know what'

Close to Becoming Irrelevant http://www.openeducation.net/2008/08/12/higher-education-dangerously-close-to-becoming-irrelevant/ today's typical college classroom is completely out of step with the business world described in Thomas Friedman's, "The World Is Flat"

Why Education is Irrelevant http://www.irrmag.com/features/therightwayblog/334-why-education-is-irrelevant.html
" the vague "virtues of education" by claiming their commitment to teaching so-called "critical thinking skills," which when you boil away all the communist overtones, becomes nothing more than questioning your betters. Just what we need in this country: MORE questioning of our leaders; more useless, pointless distractions from common sense public policy decisions"

Higher Education in Uganda?
http://www.africanexecutive.com/modules/magazine/articles.php?article=2529
Graduates in environmental management courses are still looking for jobs on the streets of Kampala. Why? They were taught how to manage the environment but not how to improve or protect it

Two-thirds of pupils say science is irrelevant
http://findarticles.com/p/articles/mi_qn4156/is_20081109/ai_n31016014/
"...the perception that science is difficult. We've pretended you have to be a genius and everyone needs to be an Einstein. There are high flyers in every discipline, but the vast majority of us aren't geniuses

## 14. Generalized environmental deterioration

Wikipedia http://en.wikipedia.org/wiki/Environmental_degradation Environmental degradation is the deterioration through depletion of resources such as air, water and soil; the destruction of ecosystems and the extinction of wildlife

Environmental conditions  http://pespmc1.vub.ac.be/ENVICOND.html  The natural world is affected by developments such as ozone depletion, deforestations, species extinction and the greenhouse effect

Population growth and environmental deterioration  http://www.awi.uni-heidelberg.de/with2/Discussion%20papers/papers_2003_2005/dp400.pdf  The complex interdependencies between demographic change, economic development and the use of the environment might be one reason why one can only find few articles in environmental economics dealing with this issues

Pollution & Environmental Deterioration http://www.ehow.com/list_7495992_problems-pollution-environmental-deterioration.html If you understand the major problems caused by pollution and environmental deterioration, you can take personal steps to slow or stop damage to the biosphere and increase the amount of usable natural resources

Environment degradation & natural capital http://www.dancewithshadows.com/society/south-africa-environment.asp Report also re-emphasizes the very direct connections between the health of the natural environment and properly functioning ecosystems on the one hand, and sustainable development, poverty alleviation and human health and well-being, on the other

## 15. Generalized lack of agreed-on alternatives to present trends

Futures of technologies http://carbon.ucdenver.edu/~bwilson/TrendsAndFutures.html Instructional designers and providers can benefit enormously by stepping back, reviewing broad trends, and forecasting likely scenarios based on those trends

Cultural transitions http://www.laceweb.org.au/bla.htm  And the alternatives to neo-colonial citizenship, and welfare institutions? In practical terms, such alternatives do not exist in concrete forms

Self-directed development http://sunsite.utk.edu/FINS/OmniCapital/Fins-OC-02.htm The conceptual plan for the settlement, which will support both a permanent and transient population, will be based on the ethic of self-realization

Differing objectives http://web.uvic.ca/hr/managertoolkit/teambuilding/team_effectiveness_critique.html

An organization is a collection of groups. The success of an organization depends on the ability of the groups within it to work together to attain commonly held objectives. Because organizations are becoming increasingly more complex, their leaders must be concerned with developing more cohesive and cooperative relationships between individuals and groups

Unresolvable disputes http://www.articlecity.com/articles/legal/article_1723.shtml The job of an arbitrator is to stay as objective as possible and provide a decision or resolution that he/she think is fair. It does not necessarily have to be advantageous for both parties. In fact, similar to litigation, there is usually a winner and a loser in arbitration

## 16. Widespread failure to stimulate man's creative capacity to confront the future

Imagining Missouri's Future
http://www.missouristate.edu/longrangeplan/imagining.htm To imagine the future, the University has a special responsibility to educate students about social goals, public purposes and values, and the ethics of citizenship as well as to encourage students to have a personal sense of responsibility for the global society.

Prototype community development http://sunsite.utk.edu/FINS/OmniCapital/Fins-OC-02.htm The conceptual plan for the settlement will support both a permanent and transient population based on the ethic of self-realization

Creative Thinking Mind Map http://blog.iqmatrix.com/mind-map/unlocking-your-creative-genius-potential-mind-map Creativity is the hidden capacity to think about ourselves, others, objects, events and circumstances in original and unique ways

Systemic Approaches to Foresight
http://www.infinitefutures.com/essays/publichealth/foresightfan.shtml futures-focused thinking provides a worldview which consistently looks at the long-range options and considers alternative possible futures which may confront our goals

Governance and Communities http://www.wfs.org/2010business.htm social mood is a primary driver of patterns in politics, entertainment, and even demographics. Understanding the socionomic model allows for probabilistic prediction of social trends and events, including macroeconomic trends

## 17. Continuing deterioration of inner-cities or slums

The Next Slum? http://www.theatlantic.com/magazine/archive/2008/03/the-next-slum/6653/ The subprime crisis is just the tip of the iceberg. Fundamental changes in American life may turn today's McMansions into tomorrow's tenements

Slums facts http://www.encyclopedia.com/topic/Slums.aspx Slums are squalid sections of a city or town, areas in which most inhabitants are in or near poverty, stores and residences are cheap and dilapidated, and streets are narrow and blighted

Global Report on Human Settlements http://www.unhabitat.org/downloads/docs/GRHS.2003.2.pdf many must draw upon internal wells of resilience just to cope each day. However, out of unhealthy, crowded and often dangerous environments can emerge cultural movements and levels of solidarity unknown in leafy suburbs

Inner cities http://www.skidmore.edu/~bturner/www.rapinnercities.html Truman's plan of Urban Renewal was the first Federal program that was committed to the restoration of cities.

Urban decay http://en.wikipedia.org/wiki/Urban_decay Urban decay is the process whereby a previously functioning city, or part of a city, falls into disrepair and decrepitude. It may feature deindustrialization, depopulation or other changes

## 18. Growing irrelevance of traditional values and continuing failure to evolve new value systems

Traditional Values Coalition http://en.wikipedia.org/wiki/Traditional_Values_Coalition a Christian Right organization that represents, by its estimate, over 43,000 Christian churches throughout the United States of America

Identity and the Metanarrative http://www.wdavidphillips.com/2008/12/08/identity-and-the-metanarrative/ metanarrative is that great overarching story that helps define our lives and place us within all of history ... when it works

Don't Succumb to Pundits http://www.huffingtonpost.com/robin-lakoff/mind-over-meta-dont-succu_b_445029.html the discussion of human events in terms of metanarrative, or narrative about narrative, in general is a relatively new explanatory systematics that has spread, over the last thirty years or so

Hypermedia and Multimedia http://edweb.sdsu.edu/Courses/EDTEC653/F-9_Main.html Media are the dominant institutions of communication and socialization of the post-modern (or information) society, as compared to the modern world in which secular institutions of education set the tone for the metanarrative

Leadership http://www.ualberta.ca/~iiqm/backissues/3_3/html/irvingklenke.html leaders who do not possess an integrated metanarrative may lack the historical-narrative context necessary to frame new meaning

## 19. Inadequate shelter and transportation

Homeless shelters http://www.highbeam.com/doc/1P2-71983.html D.C. Council member called for the establishment of an independent monitor to oversee the District's homeless system

Temporary emergency shelters http://weburbanist.com/2008/11/12/lifesaving-temporary-emergency-shelters-buildings/ With more than 35 million refugees worldwide, it's become clear that housing solutions for displaced people need to be made a priority

Carts for the Homeless http://www.dosomething.org/project/carts-homeless Los Angeles is the homeless capital of the nation; there are over 80,000+ that sleep on the streets of LA nightly

Homeless women http://www.ibiblio.org/rcip/women.html According to some estimates, between 70% and 90% of homeless families in America are headed by women. Even more significant, the population of homeless families has increased by 35% since 1989

Transportation for America disabled http://t4america.org/tag/disabled/ Transit agencies in many of our smaller communities are chronically underfunded and woefully ignored by State Departments of Transportation who are mostly concerned with using their federal transportation dollars to pour new asphalt and open new highways

## 20. Obsolete and discriminatory income distribution systems

Distribution economics http://en.wikipedia.org/wiki/Distribution_%28economics%29 refers to the way total output or income is distributed among individuals or among the factors of production

Income Distribution http://www.wiley.com/WileyCDA/WileyTitle/productCd-EHEP001023.html  The early years of the twenty-first century have witnessed a struggling middle class despite robust growth in the overall U.S. economy, yet real median family income actually fell by 3 percent

Income Distribution in Nambia
http://academic.research.microsoft.com/Paper/5390109.aspx  Within-group and between-group socioeconomic inequality in a very young and rapidly developing country

American Income http://ideas.repec.org/a/aea/aecrev/v60y1970i2p261-69.html  The impact of taxes and transfer payments on the distribution of income


## 21. Accelerating wastage and exhaustion of natural resources

Natural Resources http://www.econlib.org/library/Enc/NaturalResources.html  The *effective stocks* of natural resources are continually expanded by the same technological developments that have fueled the extraordinary growth in living standards because Innovation has increased the productivity of natural resources

Resource Exhaustion
http://library.thinkquest.org/26026/Economics/resourceexhaustion-alimit.html  The majority of economists agree that while there is no absolute limit on growth, exhaustion of natural resources could restrict future economic growth

Projected dates for the exhaustion of natural resources
http://terresacree.org/ressourcesanglais.htm  Once the peak is passed, production declines until the resource is completely exhausted. In practice, the peak is reached when about half of the resources have been extracted

Debating resources exhaustion
http://answers.yahoo.com/question/index?qid=20080428042809AAmQShO " *Like Global warming there are two sides that have an equal educational background. Make your own decision about something that important*"

Developing countries
http://nazret.com/blog/index.php/2008/06/25/ethiopia_s_natural_resources_near_exhaus  While Africa still has more biocapacity than it uses, this margin is shrinking, largely due to population growth

91

## 22. Growing environmental pollution

Pollution http://en.wikipedia.org/wiki/Pollution the introduction of contaminants into a natural environment that causes instability, disorder, harm, or discomfort to the ecosystem

India http://www.gits4u.com/envo/envo4.htm The skies over North India are seasonally filled with a thick soup of aerosol particles all along the southern edge of the Himalayas, Bangladesh and the Bay of Bengal

Global Concerns Growing https://www.greenbook.org/marketing-research.cfm/global-environmental-concerns-growing Japan lists pollution as its number one concern, followed by climate change. In comparison, pollution and climate change appear as number 11 and 15, respectively, among America's list of concerns

Images of Environmental Pollution http://images.search.yahoo.com/search/images?_adv_prop=image&fr=ytff1-sunm&va=22.+Growing+environmental+pollution

Canadian farmer sues global company http://www.guardian.co.uk/environment/2008/jan/22/pollution.gmcrops Compensation sought for compensation for farmlands "contaminated" with genetically modified material

## 23. Generalized alienation of youth

Youth suicide http://www.greenleft.org.au/node/42422 "*Most suicides in Australia go unreported by the media because of concerns the publicity could lead to further death.*" Not reporting or publicly discussing suicide means the issues behind suicide get neglected

Political alienation http://indiacurrentaffairs.org/globalisation-and-political-alienation-of-youth-r-arun-kumar/ Comprehension and sensitivity to the socio-economic problems of a society depends on that particular stage of the society and the corresponding consciousness level of the people

Violence http://www.greaterkashmir.com/news/2010/Dec/20/need-to-address-alienation-of-jk-youth-sonia-gandhi-43.asp An entire generation in India that has witnessed nothing but violence is feeling isolated

Social dislocation
http://www.oregonlive.com/opinion/index.ssf/2011/01/coming_to_grips_with_the_alie n.html Somali-American teenager by the FBI raises more questions ... young Americans who don't feel part of the fabric of our community could turn to any malevolent sub-group offering a sense of belonging ... sense of identity begins to shift from their religion, tribe or clan to an American identity based on a U.S.-centric vision of race

Double alienation http://www.usip.org/publications/double-alienation-and-muslim-youth-europe Youth feel the most strongly tied to the countries in which they live rather than ancestral homelands and struggle to "fit in"


## 24. Major disturbances of the world's physical ecology

Ecology http://en.wikipedia.org/wiki/Ecology the relation of living organisms to each other and their surroundings

Freshwater http://lakes.chebucto.org/ZOOBENTH/BENTHOS/ii.html Aquatic communities downstream of many municipalities change due to the effects of urban stormwater runoff and solid waste disposal

Landscape http://www.absoluteastronomy.com/topics/Landscape_ecology spatial structure affects organism abundance at the landscape level, as well as the behavior and functioning of the landscape as a whole

Global Warming
http://news.search.yahoo.com/search/news?p=global+warming+facts&ei=UTF-8
Statistical data covering the whole earth over years of time supports global warming. General retreat of mountain glaciers during the past century is one example.

Climate Science Divides Us / Energy Technology Unites Us
http://blogs.forbes.com/energysource/2011/01/11/why-climate-science-divides-us-but-energy-technology-unites-us/ The United Nations treaty process has devolved into an endless exercise in empty promises and angry recriminations. The growth of global carbon emissions has only accelerated in the 13 years since Kyoto was signed.

Global Warming Hoax http://www.globalwarminghoax.com/news.php Prevailing belief in the insignificance of man in the warming equation and the marginal value of experts

## 25. Generally inadequate and obsolete institutional arrangements

Is marriage as a social institution becoming obsolete
http://www.answerbag.com/q_view/188123 Blog: "The concept of marriage is adapting to the times"

An economic theory of institutional change
http://www.cato.org/pubs/journal/cj9n1/cj9n1-1.pdf Institutions can be defined as the behavioral rules that are observed by the members of a society. Institutions are human devices designed to cope with uncertainty ... and ... institutional changes are costly and often require collective action. The state will adopt a new institution only to the extent that the benefits to the state are higher than the costs. The failure of the state to institute the most efficient arrangements can occur because of ideological reasons, group interest conflicts, the limitation of social science knowledge, and so on

Collaboration across boundaries
http://www.neeea.org/NEJEE/NEJEE_Collaboration_Across_Boundaries.pdf
We will be habitually tempted to rely on the industrial lens that has shaped our institutional arrangements, and this obsolete lens will inevitably fail us and our purpose

Technological revolution
http://xroads.virginia.edu/~CLASS/am485_98/archive/x_technology/drucker5.html
Drucker: Without a shadow of doubt, major technological change creates the need for social and political innovation

Federalism
http://xroads.virginia.edu/~CLASS/am485_98/archive/x_technology/drucker5.html
Institutional trajectories and processes of change display more variation than is often assumed ... and ... there is no agreement on what are the institutions that matter

## 26. Limited understanding of what is "feasible" in the way of corrective measures.

US Dept Energy (EPA) Corrective Measures
http://homer.ornl.gov/nuclearsafety/env/guidance/rcra/study.pdf The objective of a CMS is to identify and evaluate alternative remedies and to recommend a remedy(s) for remediation

Redondo Beach, CA
http://www.redondo.org/depts/recreation/facilities/seaside_lagoon/rehabilitate/Option

%201.asp  No guarantee that a feasible corrective measure will be identified as part of the TSO findings; Perpetuates an antiquated water system design.  Five options: no tradeoff guidance.

US EPA Corrective Action Training Curriculum
http://www.epa.gov/osw/hazard/correctiveaction/curriculum/
Proposes setting training goals for new and experienced federal and state Corrective Action project managers

California guidance for corrective measures
http://www.dtsc.ca.gov/SiteCleanup/upload/Appdx_C4_083108.pdf
Respondent should include a table that summarizes the available technologies and the advantages and disadvantages of each

Water Quality http://www.extension.iastate.edu/Publications/PM1699.pdf  High nitrogen in well water is an indication of environmental contamination. Potential sources of NO3-N are leaching from agricultural fields or greenhouses, fertilizer storage tanks, or animal waste. • No **economically** feasible corrective measures available.


## 27.  Unbalanced population distribution

Population Distribution & Migration
http://www.un.org/popin/icpd/recommendations/expert/10.html migration is a rational response on the part of individuals and families to interregional differences in opportunities ... UN recommends increasing Governments' capacity to manage urban development ... It was emphasized that there was a need to integrate population distribution policies, including urbanization policies, into national development strategies.  In so doing, it was important to keep in mind that rural and urban development were two sides of the same coin.  Strategies that emphasized one at the expense of the other were doomed to failure.

Population Change http://www.fao.org/sd/wpdirect/WPan0021.htm the growth rate for Africa (2.7%) is much higher than the rate for all developing countries (1.7%) ... the overall proportion of urban population (29%) is significantly under the averages for all developing countries (38%)

UN population policy http://unjobs.org/topics/population/population-dynamics/population-policy Almost two thirds of countries, which viewed population growth as too low in 1999, have below replacement level fertility. They are primarily located in Europe

China: Population, Consumption and Social Services
http://www.acca21.org.cn/english/index.html (*website occasionally unavailable*) the extremely large population base, low levels of competence, and unbalanced coordinate distribution (Administrative Centre for China's Agenda 21 is the co-lead for NZEC together with AEA in the UK)

UNDP
http://www.thefreelibrary.com/Globalization+is+rapid+but+unbalanced.%28Statistical+Data+Included%29-a057590590 The global, professional elite enjoys a world of open borders and abundant goods, but billions of others find borders as impassable as ever ... the process is concentrating power and marginalizing the poor

## 28. Ideological fragmentation and semantic barriers to communication between individuals, groups, and nations

Cyber nations http://www.online-deliberation.net/conf2005/viewabstract.php?id=27 The tendency for groups to segregate into increasingly shared perspectives in social networks is well established ... online newsgroups currently seem to preserve a diversity of perspectives

Neo-midievailism
http://www.allacademic.com/meta/p_mla_apa_research_citation/0/7/3/3/9/p73390_index.html Globalism is leading to complex economic interdependence ... ideological fragmentation, multiculturalism, postmodernity and the expansion of diverse, ad hoc processes of global governance

Latin America http://www.nd.edu/~mcoppedg/crd/ddlaps.htm Most Latin American party systems change so often and in so many respects that the 'typical' party system of each country can be described only in imprecise terms. ... there is almost as much difference within each country as there is across the countries of Latin America

Syria http://free-syria.com/en/loadarticle.php?articleid=34839 "*While Hariri stood for the state, moderation, modernization, and political and economic integration in the world, Nasrallah was the embodiment of militant jihad, fundamentalism, ideological fragmentation and incitement*"

FaceBook blog http://www.facebook.com/notes.php?id=21647397459 "*Fox News represents part of the ideological fragmentation of news in the United States. By "ideological fragmentation" I mean that citizens who use television as a primary news*

96

*source tend to gravitate toward news sources that most closely reflect their own political views and commitments"* – attributed to Joseph Uranowski Thursday, June 17, 2010

Semantic barriers http://www.blurtit.com/q757653.html same word or symbol means different things to different individuals ... in many countries in the West, the red flag indicates danger; however, in South Korea white flag is used for the same purpose

## 29. Increasing a-social and anti-social behavior and consequent rise in criminality

Antisocial people
http://www.law.duke.edu/shell/cite.pl?69+Law+&+Contemp.+Probs.+7+%28winterspring+2006%29
Social scientists generally agree that a paradigm shift has occurred over the course of the last three decades ... away from a culturally centered, social learning causal model toward a balance of genetic and environmental factors. Yet genetic predispositions, though important, are more deleterious in the presence of adverse environments

Antisocial behavior http://www.wisegeek.com/what-is-antisocial-behavior.htm
Antisocial behavior can be generally characterized as an overall lack of adherence to the social mores and standards that allow members of a society to coexist peaceably. According to some studies, individuals with antisocial behavior disorders are responsible for about half of all crimes committed, though they make up only about five percent of the population

Anti-social causes http://uk.toluna.com/pollpagev2.aspx?pollid=44379 online survey reports 30% of respondents say "society in general"

Personality Disorders http://www.amazon.com/Antisocial-Behavior-Personality-Disorders-Hostility/dp/1573927015 'There is a growing incidence of sociopathic antisocial behavior ...coupled with an attitude of moral apathy' ... democratic ideals may have increased antisocial behavior ... education and mass media may have roles too

Explanations for Criminal Behavior
http://www.ehow.com/about_4697993_explanations-criminal-behavior.html Attitudes about wealthy defendants have also undergone a sea change following the 2007-08 economic meltdowns that cost many Americans their home values and life savings. Nearly a century after the first formal attempts to explain criminal behavior began, a consensus seems farther away than ever

## 30. Inadequate and obsolete law enforcement and correctional practices

Updating an enforcement model
http://www.meridianbooster.com/ArticleDisplay.aspx?e=2921826  greater collaboration and intelligence sharing among law enforcement agencies ... alleged     police misconduct complaints will be resolved faster and more efficiently ... integrated approach to fight gangs and organized crime ... clear performance expectations and accountability mechanisms

Privacy Law for Digital Society
http://digg.com/news/story/1986_privacy_law_inadequate_for_2011_digital_society
The mentality of law enforcement is that since there is information available about suspects, law enforcement officers should have free reign to sift through it ... if a home is searched, a search warrant specifies what can be searched

UN on drugs and crime
http://www.unodc.org/eastasiaandpacific/en/Projects/2009_08/drug_law_enforcement.html
The biggest problem with regard to drug supply reduction is the inadequate and ineffective nature of border security enforcement ... information collection and sharing procedures are inadequate

Confinement jeopardizes correction http://www.marketwire.com/press-release/Correctional-Investigator-Releases-2009-10-Annual-Report-Conditions-Confinement-Jeopardize-1348044.htm Overcrowded prisons with inadequate infrastructures... large number of mentally ill and Aboriginal inmates ... complex inmate profile that includes histories of gang membership, substance abuse and chronic illness

Limited evidence-based correctional practice
http://www.nga.org/Files/pdf/0805SENTENCERES10.PDF understand the offender, motivate the offender, select the intervention ... continuity of structure, treatment, and accountability

Juvenile programs http://www.edjj.org/focus/education/ More than 125,000 youth are in custody in nearly 3,500 public and private juvenile correctional facilities in the United States ... intense educational, mental health, medical, and social needs ... media's negative portrayal of troubled youth distorts the extent and nature of delinquency and may also erode public support for correctional education programs

## 31. Widespread unemployment and generalized underemployment

USDA reports on financial crisis impact on food
http://www.globalresearch.ca/index.php?context=va&aid=16168
In the US alone in 2008, ~17,000,000 children (22.5% lived in households in which food at times was scarce ... a 4,000,000 increase in 1 year. The number of frequently unfed youngsters rose from ~700,000 to ~1,100,000

Unemployment http://en.wikipedia.org/wiki/Unemployment Unemployment as defined by the International Labour Organization occurs when people are without jobs and they have actively looked for work within the past four weeks.

US disbelief http://sociologias-com.blogspot.com/2009/11/rising-poverty-widespread-unemployment.html "*It's frankly just deeply upsetting,*" said James D. Weill, president of the Washington-based Food and Action Center. As the economy eroded, Weill said, "*you had more and more people getting pushed closer to the cliff's edge. Then this huge storm came along and pushed them over.*"

Nicaragua http://www.indexmundi.com/nicaragua/economy_profile.html Nicaragua, the poorest country in Central America, has widespread underemployment and poverty. GDP fell by almost 3% in 2009, due to decreased export (60% as textiles/apparel) demand in the US and Central American markets. Although in 2004 Nicaragua secured some $4.5B in foreign debt reduction, the nation still carries $3.6B in debt. Unemployment >8%, Underemployment >46%. In 2005, 48% of the population was below the poverty line.

## 32. Spreading "discontent" throughout most classes of society

China http://www.bbc.co.uk/news/world-asia-pacific-12002253 Social discontent in China has risen this year, according to a top Chinese think tank. They say that China is fast moving from an agricultural society to an industrial one, with more farmers leaving the land for the cities.

Individualism in Corporate Society http://www.amazon.com/Manufacturing-Discontent-Individualism-Corporate-Society/dp/0745324061 Corporate power has a huge impact on the rights and privileges of individuals — as workers, consumers, and citizens. People perceive themselves as having choices, when in fact most peoples' options are very limited.

Venezuela http://venezuelanalysis.com/analysis/5547 The electricity cuts are one example among a handful of controversial government decisions—including a currency devaluation, the imprisonment of bankers, and the closing and expropriation of businesses accused of price speculation … and apply its nationwide electricity rationing plan to its nation's capital … merchants shut their stores in anticipation of trouble

United States http://www.amazon.com/Why-We-Hate-Discontent-Millennium/dp/0307406628 the radical social changes of the 1960s and the recent technological revolution have drastically altered the pace of life, leaving Americans morally and existentially tired, disoriented, anchorless, and defensive … causing them to to hate themselves (and each other) at a time of national prosperity

## 33. Polarization of military power and psychological impacts of the policy of deterrence

Turkey http://www.meforum.org/2160/turkey-military-catalyst-for-reform Analysts generally consider military influence in politics and society to be a critical impediment to the development of democratic political and civil rights and freedoms. Turkey may be an exception. While the Turkish constitution certainly does not endorse coups, Turkish popular distrust of politicians has generally led the public to support military action. [Without strong military role] the checks and balances of Turkish society might collapse

United States http://pinione.blogspot.com/2010/02/economic-polarization-in-america.html Five symptoms that previous world powers have experienced before they have collapsed: a sense of something was going wrong with the country; the role of religion or the excessive role of religion; economic polarization within the country; geo-political hubris; and the national debt … With millions of Americans out of work and Wall Street bankers receiving millions in bonuses, we have a vivid example of economic polarization …

Oxford Companion to Military History http://www.answers.com/topic/deterrence Deterrence is common to many human relationships and situations, ranging from raising children to controlling crime: however, the term has become best known as a feature of military strategy… In 1945, it was suggested that the psychological impact of the [nuclear] explosion might be more valuable to U.S. military objectives than the immediate physical destruction

Deterrence and Terrorism http://www.e-ir.info/?p=4208 Psychological frameworks cannot deter actors who are willing to sacrifice their lives to kill others … terrorist actions suggest that the actor who carries out violence has arrived there as a result of a logical strategic choice, given its optimal efficacy and a lack of effective alternatives

Deterrence Theory
http://www.abolishnukes.com/short_essays/deterrence_theory_whitmore.html The threat underlying nuclear deterrence is that aggressive acts will be answered by nuclear weapons retaliation. If that threat lacks credibility, then deterrence is at least uncertain and perhaps inoperative.

## 34. Fast obsolescing of political structures and processes

Investing http://nathanjensen.wustl.edu/me/files/WP2_06.pdf [Stable] institutions have an important role in attracting foreign investment and sustaining the nature of the investment environment. Predictability of economic and foreign policies have been identified as key determinants of [private] investment … on the other hand, the more Foreign Direct Investment in the economy, the more incentive there is for the host government to push forward with reforms.

Civil service http://sunzi1.lib.hku.hk/hkjo/view/50/5000141.pdf Pakistan's long honored civil service has become a cultural artifact caught in transitional time. British colonial rulers introduced progressive service changes into India in the late 1800s. Civil service officers within the imperial tradition were expected to live exemplary lives; contributing in effect to the formation of constructive habits and good character – yet managed to live aloof from the masses

Transnational corporations http://epubl.luth.se/1402-1773/2003/047/LTU-CUPP-03047-SE.pdf Corporations which enter a foreign nation eventually transfer their technology and skills to that nation, and this transfer changes the relationship between nations and transnational corporations. In the early 20th century and also just after World War II, there was a rapid growth in the number of transnational corporations (~60,000) which now impact political affairs.

Management teachings
http://www.icmrindia.org/courseware/Management%20of%20Multinational%20Corporations/Management%20of%20MNCs-Contents.htm Managing Complexity through Flexible Coordination-Characteristics of Transnational Organizations, Integrated Network, Roles and Responsibilities of Subsidiaries, Organizational Processes, Developing Transnational Managers-Business Managers, Country Managers, Functional Managers, Corporate Managers, Managing the Transnational Process.

Private utilities http://knowledge.wharton.upenn.edu/papers/1256.pdf Utilities are susceptible to opportunistic behavior by government officials as a result of low operating costs once installed ... economic or political "shock" (e.g., global financial crisis) strengthen incentives for government officials to behave opportunistically

Civil control  http://g4t.info/dienst.htm  Creative freedom is linked to military order in the sense that freedom is defined with reference to sets of rules.  One feels free when one may choose not to obey a rule that (most) others must (slavishly) obey ... If the obsolescing figure of a nation persists, it persist in the bodies of its police and soldiers"

## 35.  Irrational agricultural practices

Colombia http://www.fao.org/DOCREP/006/AD077E/AD077e17.htm  The spontaneous settlements, uncontrolled, destroy about 1,000,000 hectares of forest each year, as a result of shifting agriculture. Primitive and irrational agricultural practices bring about degradation of the soil and hydrological imbalance.

Sub-Saharan Africa  http://www.atnesa.org/challenges/challenges-havard-environment.pdf  Irrational agricultural practices, including improper use of animal traction, have often aggravated environmental degradation.

Pakistan https://tspace.library.utoronto.ca/bitstream/1807/9918/1/Ian%20Macdonald-Rationality%20and%20Risk.pdf Despite emerging appreciations of contextual knowledge systems, elements of diversity in mountain farming systems are often characterized as irrational and as obstacles to achieving the production goals of 'modernized' agriculture

Northern China http://www.ingentaconnect.com/content/ip/ooa/2003/00000032/00000003/art00005 production has been impaired by frequent natural disasters, salinization and desertification. Moreover, irrational agricultural policies and practices, poverty and other socioeconomic factors have brought about degradation

## 36.  Irresponsible use of pesticides, chemical additives, insufficiently tested drugs, fertilizers, etc.

Animal poisoning  http://www.scotland.gov.uk/News/Releases/2010/08/03125256 Pesticides were involved in almost half of the incidents (74 incidents or 47 per cent) with

36 of those categorized as the abuse of pesticides, i.e. deliberate and illegal attempts to poison animals.

Pesticidev http://www.satnet.org.ug/downloads/Agromisa%20Documents/Pesticides%20compounds%20%20use%20and%20hazards.pdf Thus, a vicious circle seemed to develop, in which more frequent applications and higher dosages were seen as the unavoidable answer to increasing pest occurrence, 'the pesticide treadmill'.

Food additives http://www.guardian.co.uk/uk/2007/sep/19/lifeandhealth.health The number of "additive- and preservative-free" products launched in the UK doubled between 2004 and 2006, from 400 to just over 800. Government agency seen as acting in the interests of the food industry rather than consumers ... concern for link between specific additives and hyperactive behaviour in young children ... need to restrict use of synthetic colours ...

Untested supplement http://www.biotivia-blog.com/2009/01/09/urgent-health-alert-warning-supplements-containing-a-compound-called-tween-are-not-fda-approved-and-may-be-harmful-to-humans The company claims that this chemical, also known by the trade name Tween and Tween 80, increases bioavailability of the resveratrol contained in the company's supplement. There is absolutely no evidence to support this claim.

Lawn care blog http://forums2.gardenweb.com/forums/load/lawns/msg04153641567.html .. *In some areas, laws have been passed banning the use of fertilizers with Phosphorus except when reseeding or when soil tests prove the soil is deficient*

## 37. Growing use of distorted information to influence and manipulate people

Propaganda http://en.wikipedia.org/wiki/News_propaganda News propaganda is covert propaganda packaged as credible news without transparency as to source and motivation

Deliberately-distorted news http://www.newsmax.com/Manage/Videos/VideoGallery/Jimmy-Carter--FNC-Commentators--Deliberately-Disto Video interview with President Jimmy Carter, commenting on his relationship with the media

Public relations http://www.psrast.org/propaganda.htm An efficient trick to manipulate people is to associate the product with issues that they feel are especially important or have a high status. Therefore, the propaganda has addressed topics like world hunger, cure of serious diseases and High Tech status of the technology.

Fact checking http://factcheck.org/ We are a nonpartisan, nonprofit "consumer advocate" for voters that aims to reduce the level of deception and confusion in U.S. politics. We monitor the factual accuracy of what is said by major U.S. political players in the form of TV ads, debates, speeches, interviews and news releases

Center for Media and Democracy http://www.prwatch.org/cmd Supports strategic public education campaigns, educate the public about financial reforms and health insurance industry, response to the Supreme Court's decision in the Citizens United case expanding corporate "rights," Defend The Press campaign with the National Press Club

Markets for information http://www.econ.ucsb.edu/papers/wp34-98.pdf When can reputation be an effective constraint on deliberate misinformation? ... frequency of truthfulness is 81% when the asset value is low

Information literacy http://www.ics.heacademy.ac.uk/italics/vol5-1/webpages/Whitworth_final.htm What characteristics does a statement possess which encourage hearers or readers to interpret it as a seriously-meant contribution to discussion? Among the things which can drag public discourse away from Habermas's ideals is distorted information.

Thailand http://blogs.reuters.com/andrew-marshall/2010/08/03/thailand-the-queen-and-cnn/ A letter from Her Majesty Queen Sirikit of Thailand praising a young woman who wrote a critique of foreign media coverage of the Thai political crisis has given a rare insight into the monarchy's views on the country's troubles

Distorted Thinking http://www.sandf.org/articles/ThinkTrap.asp a list of distorted thinking patterns, some of which are drawn from the fields of cognitive/behavioral therapy ...

## 38. Fragmented International monetary system

International monetary systems
http://en.wikipedia.org/wiki/International_monetary_systems A system within which nations agree to fixed but adjustable exchange rates where the currencies are pegged against the dollar, with the dollar itself convertible into gold

International monetary systems
http://www.tutorgig.com/ed/International_Monetary_Systems International monetary systems are sets of internationally agreed rules, conventions and supporting institutions that facilitate international trade, cross border investment and generally the reallocation of capital between nation states.

China
http://www.chinasecurity.us/index.php?option=com_content&view=article&id=309&Itemid=8 The global economic crisis has ignited concerns about the functioning of the international currency system. The value of China's massive foreign exchange reserves, the fortunes of Chinese exporters and the flows of hot money into the country are all shaped by the USD/RMB exchange rate .. and China's heavy reliance on trade-related growth

Orderly decline of the dollar http://www.zerohedge.com/article/george-soros-united-states-must-stop-resisting-orderly-decline-dollar-coming-global-currency?+the+survival+rate+for+everyone+drops+to+zero%29 The international banking elite very much want a global currency and a "New World Order" ... the Council on Foreign Relations, the Trilateral Commission or the Bilderberg Group ... weakening a currency and having moderate levels of inflation can improve the competitiveness of an economy ... Passive savers are penalized, active investors are rewarded. The UK, Sweden, Korea and Poland are all recent examples of that concept in action

Significance of the Euro http://www.highbeam.com/doc/1G1-118987710.html The monetary integration that took place at different times within the nations of Europe is now taking place between the nations of Europe. Monetary integration has a sense of inevitability because it is enormously more efficient. There is, however, a sense in which the European Monetary Union departs from historical tradition. Money has a cultural dimension; it has been called the centerpiece of civilization. The emergence of the Euro represents a trend toward integration.

## 39. Growing technological gaps and lags between developed and developing areas

Information technology http://en.wikipedia.org/wiki/Digital_gap The digital divide is the gap between people with effective access to digital and information technology and those with very limited or no access at all. It includes the imbalance both in physical access to technology and the resources and skills needed to effectively participate

Space Technology http://www.un.org/News/Press/docs/1996/19961111.gaspd93.html The Chairman of the United Nations Committee on the Peaceful Uses of Outer Space

105

stated that current uses of space technology had only begun to indicate how such technology could improve the human condition, particularly in developing countries … enhancing science and technology in developing countries … legal matters pertaining to the geostationary orbit …

Invention Technology
http://www.unctad.org/Templates/Page.asp?intItemID=3796&lang=1 The technology gap exists between those who can create and innovate to produce new technologies and those who cannot. It also exists between those who can access, adapt, master and use existing technologies and those who cannot. In Japan there are 861 patents granted per million people. In many developing countries the number is 0

Healthcare technology
http://gateway.nlm.nih.gov/MeetingAbstracts/ma?f=102194208.html promoting health and developing advanced medical technologies … models describe the decision-making process. Some suggest dissemination of new technologies while others suggest crystallizing the medical technology management policy. These models do not offer a suitable answer for developing nations

Global picture
http://www.caricom.org/jsp/speeches/world_bank_staff_exchange_carrington.jsp …(sometimes site is unreachable) Five (5) critical gaps between developed and developing countries … impacts advance social, cultural and technological development; encourage activities

## 40. New modes of localized warfare

New model of warfare http://powerandcontrol.blogspot.com/2006/09/new-model-warfare.html The old model of warfare - conquering and occupying territory - is not viable in the face of protracted guerilla warfare. The guerillas, if sufficiently dedicated, can wear out the occupying power. The new model is to let the insurgents do what they do not do well -- control territory in the face of a counter insurgency. The essence of guerilla warfare is mobility. Once the guerillas lose that, one of their major advantages is gone.

Asymmetric warfare http://en.wikipedia.org/wiki/Asymmetric_warfare Asymmetric warfare is war between belligerents whose relative military power differs significantly, or whose strategy or tactics differ significantly.

Cheap and easy warfare http://www.worldwidewords.org/turnsofphrase/tp-asy2.htm ... extend the idea to any military situation in which a technically weaker opponent is able to gain an advantage through relatively simple means. An obvious example is the landmine — cheap and easy to distribute, but difficult to counter.

Globalization and Asymmetrical Warfare http://www.au.af.mil/au/awc/awcgate/acsc/02-053.pdf The negative effects of globalization have continued to create a large disenfranchised population primarily centered in the Middle East, Africa, and Asia ... a changing role for information warfare ... international connectivity, access to the internet and to space-based communications, and reconnaissance capabilities ... global news coverage will focus on areas impacted by globalization ... the U.S. will lose its hold on what some have called its "media howitzers"

Israel: blog http://blog.standforisrael.org/issues/terrorism/asymmetrical-warfare Israel's challenge in fighting terrorists is a classic case of Asymmetrical warfare ... horrible twists to "guerilla warfare" by targeting civilians rather than the military, and by using their own civilian populations as human shields ... Defenses include cultivating superior intelligence networks ... redrawing an urban battlefield lessens the incentives terrorists have to operate from within population centers

## 41. Inadequate participation of people at large in public decisions

South Africa Environment http://www.docstoc.com/docs/19454581/the-role-of-public-participation-in-the-decision-making-process-of Discussions of science and technical issues dominate social issues ... no agreed definition of how to include the voices of marginalized groups ... pollutants from noxious industries in apartheid-planned communities ... resistance to industrial expansion ... rubber stamp formality ... corporatization of public participation

Evaluation frameworks http://www.rff.org/documents/RFF-DP-99-06.pdf public hearings, notice and comment procedures, and advisory committees--as well as those considered more innovative--such as regulatory negotiations, mediations, and citizen juries ... goals are: educating the public, incorporating public values and knowledge into decision-making, building trust, reducing conflict, and assuring cost-effective decision-making ...

Water resource management http://unu.edu/unupress/sample-chapters/EnhancingParticipation.pdf The world faces a growing water crisis ... Critical knowledge is distributed across governments, non-governmental organizations and water users ... conditions facilitate or hinder public involvement, and contextual factors limit the transfer of experiences … costly investment in watershed participatory management activities … even with the best of intentions, governments face daunting challenges to include participatory management …

Social Justice
http://www.planning.org/certification/examprep/pdf/socialjusticepresentation.pdf
(American Institute of Certified Planners / American Planning Assn) builds coalitions … Saul Alinsky / Sherry Arnstein / Paul Davidoff / Norman Krumholtz … disorganize in order to reorganize … the organization has to be an educational mechanism … community coalitions need to develop their own agendas … plural planning requires advocates … the planners isn't a value-neutral technician … planning commissions are political … provide more choices to those who have few …

Quality Management http://pdc.ceu.hu/archive/00002636/01/spahic.pdf  Bosnia and Herzegovina seek to harmonize with the European Charter on Self-governance … Public debates gatherings are rarely used decision-making, although prescribed by law … insufficient knowledge of modern methods of citizen participation in decision-making … 36% of surveyed citizens know their local representatives … municipalities that have introduced ISO 9001/2001quality management standards engaged three fold more citizens in public deliberations

## 42. Unimaginative conceptions of world-order and of the rule of law

The Rule of Law http://www.uiowa.edu/ifdebook/faq/Rule_of_Law.shtml a legal-political regime under which the law restrains the government by promoting certain liberties and creating order and predictability regarding how a country functions … a system that attempts to protect the rights of citizens from arbitrary and abusive use of government power … yet governments are often compelled to prioritize one goal over another to resolve conflicts in a way that reflects society's political choices

Law http://en.wikipedia.org/wiki/The_Concept_of_Law a theory of legal positivism -- the view that laws are rules made by human beings and that there is no inherent or necessary connection between law and morality.  Alternatively, a jurisprudential concept that holds that law is command backed by threat and is meant to be ubiquitous in its application.

United Nations rule of law http://unrol.org/article.aspx?article_id=132 *"strengthening the rule of law is central to achieving the vision of the United Nations for a just, secure and peaceful world"* importance of empowering individuals and the civil society, particularly the poor and most vulnerable, as well as those affected by conflict and crisis.

Countries in which civil society initiatives to strengthen the rule of law receive financial support from the United Nations include Afghanistan, China, Ethiopia, Kazakhstan, Morocco and Sierra Leone

Manifesto for an anarco-communist alternative http://news.infoshop.org/article.php?story=20051031203956348 … oppose liberal capitalism, founded upon an autonomous market regulation and which pretend to be "democratic", whereas it is based upon in essence anti-democratic ways of production and that it is wholly turned towards the realization of ruling classes' profits. We oppose State capitalism, even when it pretends to be "socialist" or "communist", whereas it is based upon a tyrannical exploitation …

## 43. Irrational distribution of industry supported by policies that will strengthen the current patterns

Biofuels http://www.accenture.com/us-en/Pages/insight-biofuels-study-part-two-summary.aspx a global industry is emerging around the production, processing and distribution of biofuels … conditions reflect excess and stranded capacity in the United States, Europe and Brazil; high and volatile feedstock prices; and low product prices … environmental benefits must be clear for motorists and business-to-business demand to support the growth of biofuels

Healthcare http://www.reuters.com/article/2009/11/06/us-usa-healthcare-doctors-special-idUSTRE5A50EB20091106 Nowhere in the United States has more doctors at its beck and call than White Plains, New York. Nearly 3,000 miles away, scaring up a doctor in Bakersfield, situated in California's economically battered Central Valley, is a lot harder. Two decades worth of U.S. healthcare data analyzed by Dartmouth Medical School at Reuters' request shows that such regional disparities are increasingly creating a nation of health-care haves and have nots.

Manufacturing http://www.acca21.org.cn/chnwp12.html Though the overall level of industrial development in China is very low, industry has become the dominant force in the national economy … However, resource allocation is quite poor, and the overall level of industrial technology is very low, resulting in the waste of resources, pollution of the environment and a weak capacity for sustainable development.

River basins http://www.chinadialogue.net/article/show/single/en/869 Irrational distribution of industry in the Yangtze River basin is to blame for the high rate of accidents that lead to water pollution. Half of China's 20,000 petrochemical companies are located in the Yangtze River basin.

## 44. Growing tendency to be satisfied with technological solutions for every type of problem

Disease http://abcnews.go.com/Technology/AheadoftheCurve/simple-solutions-fix-global-problems-hurricanes-global-warming/story?id=8890846 Steven Levitt, the authors of "SuperFreakonomics," say that throughout history, "cheap and simple [technology] fixes" often solved the world's biggest problems. Polio could have been addressed with more hospitals or more efficient iron lungs that would make breathing easier. A vaccine changed everything.

Policing http://www.gerrymcgovern.com/nt/2004/nt_2004_12_06_technology.htm Policing in the UK began to go wrong in 1966 when a policy was established where policing focused more on technology and less on people. The remedy was to get back on the streets, and start rebuilding relationships and trust with the public.

Education http://privateschool.about.com/od/educationaltechnology/f/solution.htm Have teachers transformed their teaching with technology? The reality is that technology has improved communications and access to information. That's a good thing. The downside is that we are not teaching our students how to use technology to make a living in a highly competitive global arena.

Environment http://fiesta.bren.ucsb.edu/~delmas/webpage/Papers/Delmas-Marcus-Bpol.pdf Monitoring for regulatory compliance is accomplished at the process level -- if the technology is in place, the firm complies. Alternatively the monitoring can be done ... on the actual discharge levels. Regulatory agencies typically prefer the technological solutions since these eliminate the need to monitor.

Agriculture http://www.edenfarmanimalsanctuary.org/factory_farming Society's practical rebuttal of Deep Ecology is clearly shown by society's support for factory farming. Factory farming has a materialistic, anthropocentric, consumer-oriented attitude to animals. Factory farming is part of the fibre of shallow ecology: using nature's resources for unregulated human growth, relying on technological solutions to fix socio-economic problems.

## 45. Obsolete system of world trade

Emissions Trading http://euobserver.com/9/30779/?rk=1 The economic crisis, which has shut down manufacturers, idled factories and left lorries and ships with fewer products to transport, has rendered the EU's flagship climate change policy, the emissions trading scheme "obsolete"

Oceanic Shipping
http://www.unctad.org/Templates/Webflyer.asp?docID=14175&intItemID=1528&lang=1
Maritime transport is the single most important transport mode; it has around 80 per cent of the market share in the international movement of goods. However, in some developing countries this percentage is much higher, due to cumbersome cross-border procedures and an underdeveloped land transport infrastructure. Seaborne trade in bauxite and alumina - key components in aluminium used primarily in the transport and construction industries -suffered a 23.2% decline. Phosphate rock, used as a fertilizer by the agriculture industry, suffered a 38.7% decline.

Regional Councils
http://www.columbia.edu/~jid2106/td/archives/2007/04/is_the_mfn_prin.html The Atlantic Council showed that it was no longer possible to make meaningful progress in a global negotiating system that operated through consensus.  Stuart Eizenstat, a former undersecretary of commerce in the Clinton administration *The world is moving too fast for this kind of consensus-driven, five, six, seven, eight-year rounds."*

Trade Policy http://www.thirdworldtraveler.com/WTO_MAI/ForTheirEyesOnly.html challenging balance between two potentially conflicting priorities: promoting trade expansion versus protecting the regulatory rights of governments … The World Trade Organization's "Necessity Test" places authority with the GATS Disputes Panel to determine if a law or regulation is 'more burdensome than necessary' … The Necessity Test was applied when a Canadian company claimed that California's ban of the gasoline additive MBTE [found in water supplies] failed the test.

WTO reform https://litigation-essentials.lexisnexis.com/webcd/app?action=DocumentDisplay&crawlid=1&doctype=cite&docid=18+Emory+Int%27l+L.+Rev.+421&srctype=smi&srcid=3B15&key=f9c6cdb5104e44a384ed2852006e163a In contrast to the GATT, the Agreement Establishing the WTO  has created an international governmental organization and has endowed it with full legal capacity and all attributes thereof. This includes provisions on privileges and immunities for its officials, a fully fledged organizational structure, which encompasses the position of the Secretary General, the Secretariat, and numerous other specialized bodies, as well as the capacity to establish legal relationships and to conclude agreements with other governmental and non-governmental organizations

## 46. Ill-conceived use of international agencies for national or sectoral ends

International Law http://www.biu.ac.il/SOC/besa/docs/perspectives73.pdf The antagonists in the struggle are employing the weapon of their adversaries – the rule of law – in a strategy called "Lawfare" which involves the misuse of the law to achieve objectives that cannot be achieved militarily.

Medical research agencies http://www.chrysotile.com/data/IBE%20Manuscript.pdf .Confusion and indiscriminate use of 'hazard' and 'risk' mean that the hazard data are commonly misrepresented as risk data. A common political response is to push regulatory action to extremes, citing the Precautionary Principle.

USAID conceals misuse of funds http://kabulpress.org/my/spip.php?article4763 U.S. Agency for International Development (USAID) in Afghanistan decided to illegally (i.e., without competitive bidding) to extend an existing contract. This type of award is called a "sweetheart" contract in America because the recipient has close political ties to the politicians currently in power in Washington, D.C. While USAID later announced that it was canceling the award of the contract, it refused to fire the agency officials responsible.

Misuse of Humanitarian Relief Goods http://www.doctorswithoutborders.org/publications/article.cfm?id=1465 Médecins Sans Frontières (MSF) is calling on the relevant bodies of the UN and its individual member states to take all measures needed to ensure that those suspected of having been involved in the genocide in Rwanda are brought to justice; that refugees get adequate protection and do not have to live in fear of their lives; that they get equal access to humanitarian relief goods; that the militia and military are separated from the refugees into different camps; and that adequate measures are taken to improve the security situation of refugees and of relief workers.

## 47. Insufficient authority of international agencies

AUFM limits military http://www.washingtonpost.com/wp-dyn/content/article/2010/11/25/AR2010112503211.html The Authorization for Use of Military Force (AUMF) provides insufficient authority for our military and intelligence personnel to conduct counterterrorism operations today and inadequate protections for those targeted or detained, including U.S. citizens

Revenues and Customs Authority untouchable http://www.erca.gov.et/about.php In Ethiopia, the responsibility to collect revenue for the federal government rests with the Ethiopian Revenue and Customs Authority (Hereafter cited as the Authority). In addition to raising revenue, the Authority is responsible to facilitate the legitimate movement of people and goods across the border. Simultaneously, the Authority focuses on those people and vehicles that may involve in the act of smuggling

World Bank seeks policy change at Nigerian Port Authority http://www.otal.com/nigeria/index.htm Nigerian Maritime Authority: This government parastatal is involved in protecting the country's maritime interest. Consignee's do not pay any amount to NMA during the clearing process as all NMA charges are paid by shipping lines that call Nigerian ports. The increase of NMA charges in 2003 led to the introduction of the freight tax surcharge added onto all Nigerian freights. The Federal Government announced in Dec 2004 that all NMA charges payable by shipping companies will be scrapped as from Jan 2005, following recommendations from the World Bank. However, this policy is yet to be implemented.

IMF seeks to improve Securities Market http://www.imf.org/external/pubs/ft/survey/so/2008/RES02408A.htm In many countries, particularly low-income ones, a combination of factors has undermined the capacity of regulators to regulate effectively. These include insufficient legal authority, as well as a lack of resources, skills, and political will.

USAID monitors disasters in Pakistan http://www.usaid.gov/our_work/humanitarian_assistance/disaster_assistance /countries/pakistan/template/index.html Several U.N. agencies and non-governmental organizations in Pakistan seek to coordinate responses, yet as the provincial government had used resources to respond to recent flooding, remaining resources were insufficient to address the humanitarian needs

## 48. Irrational practices in resource investment

Land grab http://www.iied.org/legal-tools/trends-natural-resource-investment-africa Land is central to livelihoods, food security, even identity — the natural result of a direct dependence on agriculture and natural resources. A recent wave of large-scale land acquisitions in poorer countries has sparked a major debate. Richer countries are buying or leasing large tracts of farmland for agricultural investment in Africa, Central and Southeast Asia, Latin America and Eastern Europe -- some deals involving hundreds of thousands of hectares.  These investments can either create new opportunities to improve local living standards, or further marginalize the poor.

Soil degradation http://www.fao.org/docrep/004/y1796e/y1796e02.htm Soil degradation is a major global environmental problem, causing widespread and serious impacts on water quality, biodiversity and the emission of climate changing greenhouse gases.  Yet a primary cause of soil degradation is the depletion of soils which results from the farming systems chosen by the farmers. Decisions such as where and when to produce, the types of techniques used - particularly in land preparation - and the level and timing of inputs

Petrochemicals http://www.mrglobalization.com/environment/233-natural-resource-trade-and-investment In the past, an important aspect of Western colonialism was investment in natural resources.  This investment then declined in importance with decolonization and the creation of the Organization of the Petroleum Exporting Countries (OPEC).  However, over the past decade or so, with rising demand from China and other major emerging economies, there has been strong growth in trade and investment in natural resources.

Foreign Direct Investment (FDI)  http://pagerankstudio.com/Blog/2011/01/what-is-resource-seeking-investment/
Resource-seeking investment is one of the types of foreign direct investment; it mainly focuses on rich raw materials, low-cost unskilled and skilled labor, technological assets, and physical infrastructure. According to the United Nations Conference on Trade and Development (UNCTAD), resource-seeking investment was the most common foreign direct investment (FDI) type in the 19th and 20th centuries

Education http://tanzaniafpsrh.blogspot.com/2010/06/from-our-media.html The Feminist Activists Coalition (FEMACT) published a protest statement commenting on the speech by Jakaya Kikwete who, a week before, said it was only foolhardy on the part of the girls themselves when they become pregnant when they are at their teen. He said the girls were the ones to blame for rushing into sex. FemACT demands the government look at kid pregnancy problem as caused by policy failure and poor resource investment in education sector.

Decentralization http://www.ciesin.org/decentralization/English/CaseStudies/Nepal2.pdf In a newly established democracy, political parties do not have extensive organizational roots at local levels. Use of subnational bodies for political gains is an expected possibility and became a source of political conflict in Nepal. When viewed in political light, decentralization is seen as a tool to engineer electoral gains. The ruling party at the centre finds it imperative to see who is in the majority at the local levels before decisions on devolution or resource allocation are made.

Global Quality Standards ISO 14021 http://www.ecodesign-company.com/documents/BestPracticeISO14021.pdf Irrational resource consumption together with irresponsible environmental pollution resulting from the entire product life cycle – raw material acquisition, manufacturing, distribution, use and disposal – is the main cause of exceeding the global environmental carrying capacity. This is because our industrial structure and consumption patterns are not designed to minimize environmental impacts. This has created a growing concern that a sustainable society may not be achievable.

## 49. Insufficient understanding of Continuous Critical Problems, of their nature, their interactions, and of the future consequences both they and current solutions to them are generating

"Wicked problem" http://en.wikipedia.org/wiki/Wicked_problem  "Wicked problem" is a phrase originally used in social planning to describe a problem that is difficult or impossible to solve because of incomplete, contradictory, and changing requirements that are often difficult to recognize

Social "messes" http://www.nautilus.org/gps/solving/wicked-problems  complex problems that change when you apply a solution. Morphological analysis is a general method for structuring and analyzing complex problem fields which 1) are inherently non-quantifiable; 2) contain non-resolvable uncertainties (both antagonistic and non-specified uncertainty); and 3) cannot be causally modeled or simulated in a meaningful way.  In his book from 1974 titled "Redesigning the Future", the operational analyst Russell Ackoff defined three levels of complex problems. The first level he called a mess; the next level he called a problem; and the third he called a puzzle

Reductive thinking http://changeorder.typepad.com/weblog/wicked_problems/index.html  Any type of reductive thinking will actually worsen a wicked problem. Being a designer can feel painful, but the real pain begins when the planning becomes reality, and the tangled mess of sticky complexity that our clients are struggling with begins to unravel in finite degrees through brute-force effort. We can only ease that pain slightly, and help them bear it as they aim to create a meaningful impact.

The Problematique http://isss.org/projects/problematique  only an effort which strives to go beyond "conventional wisdom" and methodological orthodoxy can allow us to perceive the complex dimensions of the problematique of our age, and thus set the stage for the formulation and development of the long-term options and alternative outlooks needed for policy-making

115

"Global Problematique"
http://www.globalproblematique.net/definitions/worldproblematique.html  The Global or World Problematique can be defined as the combination of: 1) the interconnected and interacting issues of all types, which affect the becoming of life in the global ecosystem, or 2) the uncontrollable and interacting circumstances of all types, which constrain the timely and effective resolution of these issues.

Attempted summary http://www.huffingtonpost.com/david-roberts/the-global-problematique_b_45070.html  I just consider myself a progressive and a humanist -- one who thinks the issues of climate change and energy depletion are not being taken seriously enough. At risk of becoming the guy in the corner ranting with the homemade cardboard sign, I'll just try to summarize what I see as a mutually reinforcing network of oncoming problems, and leave it at that

Sustainable Development http://portal.unesco.org/en/ev.php-url_id=3288&url_do=do_topic&url_section=201.html UNESCO launched the Global Problematique Education Network Initiative, otherwise known as GENIe.  GENIe is a network of universities whose fundamental objective is to provide an integrated multidisciplinary (scientific and humanistic) education, in a university context, on global issues regarding humankind and nature (population, global warming, carrying capacity, famine, poverty, etc.) with a view to establishing a sustainable world.

# Epilogue

"It is not down on any map. True places never are."

Herman Melville

Ten years after the founding of the Club of Rome the "Human Ecological Footprint" breeched the thresholds of tolerance of the carrying capacity of the earth's natural systems. It was an unprecedented historical event.

The population of the earth was 4.5 billion

Henceforth, it would take more than one earth to provide for the growing human population

Seemingly in response to the event itself Hans Jonas wrote: *The altered, always enlarged nature of human action, with the magnitude and novelty of its works and their impact on man's global future, raises moral issues for which past ethics, geared to the direct dealings of man with his fellowman within narrow horizons of space and time, has left us unprepared.* (The Imperative of Responsibility, Jonas, 1984)

Karl Jaspers, informed by the work of Max Weber, recognized that in the first millennium BCE there emerged across the known world new religious and philosophical expressions somewhat common in their characteristics despite their apparent isolation. From Confucianism and Toaism in China to Jainism and Buddhism in India, to Zoroastrianism in Persia, the Hebraic prophets and the Greeks literary and philosophical traditions there appeared an adaptive emergence in consciousness that produced a new sense of what it meant to be human. He called it the "Axial Age"

This Axial Age also produced the ethical basis between man and man to which Jonas refers. However, it did not produce an ethic between Man and the natural world on which he depends or between the present and future generations. In fact, Hannah Arendt has suggested that the Greeks invented history so that they could participate in the same eternity that they saw in Nature.

For the Greeks, nature's carrying capacities were presumed to be infinite. This attitude we still share and among its manifestations is the fact that we do not value natural systems assets or their services in our current economic planning models.

It is this asymmetrical relationship between our recalcitrant cultural habits and the new imperatives of responsibility that threaten our survival. While this condition is threatening, unlike those who endured and created the Axial Age, at least we can see more of what is happening.

I am beginning this epilogue with modern insights which predate the Club of Rome by about 20 years because they suggest the ethos that produced the Club of Rome. Out of this period and its rich reflections may come the new ideational cultural synthesis to which Pitirim Sorokin alludes below. However, whether or not this transition to a new culture is axial in proportion will be dependent on its affinity toward engaging the ethical imperatives raised by Jonas

If this transition is axial it will be in no small measure due to the contributions of the Club of Rome

## Contextual Perspectives

*"Every important aspect of the life, organization and the culture of Western society is in extraordinary crisis…Its body and mind are sick and there is hardly a spot on its body which is not sore, nor any nervous fiber which functions soundly…We are seemingly between two epochs: the dying Sensate culture of our magnificent yesterday and the coming Ideational age of the creative tomorrow." We are living, thinking, and acting at the end of a brilliant six hundred year long Sensate day. The oblique rays of the sun still illumine the glory of the passing epoch. But the light is fading, and in the deepening shadows it becomes more and more difficult to see clearly and to orient ourselves safely in the confusions of the twilight. The night of the transitory period begins to loom before us, with its nightmares, frightening shadows, and heart rendering horrors. Beyond it, however, the dawn of a new great Ideational culture is probably waiting to greet the men of the future."*

(Pitirim Sorokin, Social and Cultural Dynamics, 1937

*"We overstep history when man becomes present to us in his most exalted works, through which he has been able as it were to catch history in motion, and has rendered it communicable. What has here been done by men, who allowed themselves to be absorbed by the eternal truth, which became language through them, although it wears an historical garb, is above and beyond history and leads us along the route that passes over the world of history into that which is prior to all history and becomes language through history. In this realm there is no longer any question of whence and whither, of future and progress, but in time there is something that is no longer solely time and which comes to us above all time, as Being itself…*

*In the contemplation of the great – in the provinces of creation action and thought – history shines forth as the everlasting present. It no longer satisfies curiosity but becomes an invigorating force."*

(Karl Jaspers: The Origin and Goal of History, 1953)

The heralds of change also spoke with mytho-prophetic voices. Eiseley was an anthropologist and paleontologist by profession and a poet by nature. He may be the twentieth century's Dante in his ability to characterize the human experience in the cosmology of his own age. Nevertheless, his characterization of the "coming of man" as a phenomenon of geological proportion in terms of the consequences for our planetary home is compelling. It is  arguably expressive of a consciousness that has given us Hans Jonas' "The Imperative of Responsibility."

*"It is with the coming of man that a vast hole seems to open in nature, a vast black whirlpool, spinning faster and faster, consuming flesh, stones, soil,  minerals, sucking down the lightening, wrenching power from the atom, until the ancient sounds of nature are drowned in the cacophony of something that is no longer nature, something instead which is loose and knocking at the world's heart, something demonic and no longer planned – escaped It may be – spewed out of nature , contending in a final giant's game against its master."*

(Loren Eiseley; "The Firmament of Time, 1960)

Of the ethos from which the Club of Rome emerged, scholars will for centuries debate. Nevertheless in the second half of the twentieth century there emerged in our conscious, unconscious, and not-yet-conscious ways of knowing a most disturbing new dimension.

I have offered reflections from Pitirim Sorokin and Karl Jaspers on the "crisis of our age" and of the ontological capacities in us to engage the looming "nightmares", "frightening shadows" and "heart rendering horrors" of which Sorokin speaks because it is expresses something of the physical world Aurelio Peccei saw and Hasan Ozbekhan tried so urgently to articulate in a way that something might be done about it. The origin of the monograph "The Predicament of Mankind" is seen in Sorokin' s "Heart Rendering Horrors" (1937) and in the emerging empathy and outrage that find expression in  Franz Fanon's "The Wretched of the Earth" (1963) and Paolo Friere's "Pedagogy of the Oppressed" (1970).

However, I chose the short but dramatic warning from Loren Eiseley because it anticipates what was ultimately revealed in the "Limits to Growth" and places us not only in the context of a changing culture but also in an historical moment when it would appear that in the words of the historian

119

William Irwin Thompson, "*the whole meaning of nature, self, and civilization is overturned in a re-visioning of history as important as any technological innovation.*"

---

## In Pursuit of The Club of Rome

At the moment of its birth, the splitting of the Club of Rome may prove to be one of the great tragedies of the 20th century. The split separated the emerging practice of Systems Dynamics (the tool which gave us the "Limits to Growth") and the philosophical vision of Systems Design Science (the foundation of Structured Dialogic Design). This split separated the extrapolated mechanical insights into the future expressed in the "Limits to Growth" from the transformational social complexities that create the future. The rush to embrace a mechanical view of the world delayed engagement of the social complexities reshaping the world.

In the twenty-year follow up on "The Limits to Growth" the MIT economists Dennis and Donella Meadows and Jorgen Randers wrote a reflection for "In Context" which they entitled "Beyond the Limits to Growth". It contains a most profound confession:

> "*A perceptive teacher, watching his students react to the idea that there are limits, once wrote:*
>
>> *When most of us are presented with the ultimata of potential disaster, when we hear that we "must" choose some form of planned stability, when we face the "necessity" of a designed sustainable state, we are being bereaved, whether or not we fully realize it. When cast upon our own resources in this way we feel, we intuit, a kind of cosmic loneliness that we could not have foreseen. We become orphans. We no longer see ourselves as children of a cosmic order or the beneficiaries of the historical process. Limits to growth denies all that. It tell us, perhaps for the first time in our experience, that the only plan must be our own. With one stroke it strips us of the assurance offered by past forms of Providence and progress, and with another it thrusts into our reluctant hands the responsibility for the future. (Vargish, 1980)*
>
> *We went through that entire emotional sequence - grief, loneliness, reluctant responsibility - when we worked on The Club of Rome project twenty years ago. Many other people, through many other kinds of formative events, have gone through a similar sequence. It can be survived.*"

*"It can even open up new horizons and suggest exciting futures. Those futures will never come to be, however, until the world as a whole turns to face them. The ideas of limits, sustainability, sufficiency, equity, and efficiency are not barriers, not obstacles, not threats. They are guides to a new world. Sustainability, not better weapons or struggles for power or material accumulation, is the ultimate challenge to the energy and creativity of the human race."*

*"We think the human race is up to the challenge."*

(In Context: "Beyond The Limits To Growth;" summer 1992)

Let's focus on what is being said here:

We are at the end of an age whose "past forms of Providence and progress", whose gods and metaphysical structures, no longer work. In the same year, in the same city, and on the same campus that the authors began their work, a young MIT historian William Irwin Thompson anticipating the same sense of cosmic loneliness wrote:

*"It would still seem that we are at one of those moments when the whole meaning of nature, self and civilization is overturned in a re-visioning of history as important as any technological innovation….We will have to come right up to the edge to find out where we are, and who we are. At the edge of history, history itself can no longer help us, and only myth remains equal to the task. What we know is less than what we are, and so the politics of miracle must be unacceptable of our knowledge to be worthy of our Being."*

(William Irwin Thompson, At the Edge of History, 1971)

Sobered and matured, Dennis Meadows and Jorgen Randers in their 30 year update and republication of the Limits to Growth confessed that Donella had remained an optimist about humanity's capacity to change and Jorgen had become a cynic believing that humanity would continue to pursue short term interests while Dennis was somewhere in between. Their conclusion was:

*"Sadly, we believe the world will experience overshoot and collapse in global resource use and emissions much the same way as the dot.com "bubble" – though on a much longer time scale. The growth phase will be welcomed and celebrated even long after it has moved into unsustainable territory (this we know because it has already happened). The collapse will happen very rapidly much to everyone's surprise. And once it has lasted for*

121

*some years it will become increasingly obvious that the situation before the collapse was totally unsustainable."*

Essentially Meadows and Randers are resigned to the notion that overshoot and collapse is not only the result of the human ecological footprint *"exceeding the carrying capacities of the natural systems of the planet but also the result of the recalcitrance of world views locked in institutional structures that persist in "doing the wrong things righter and righter,"* despite the evidence of their obsolescence.

Implicit in their observation is the conclusion Hasan Ozbekhan and Alexander Christakis reached when they processed the 49 enormous world problems of the Problematique and concluded that the path to a sustainable future was – at its root – value-based.

---

**Wharton and Convergence**

In the years 1992 and 1993 Peter Drucker, the internationally recognized management consultant and social philosopher, published the "The New Realities" and "Post-Capitalist Society" in which given his temperament and discipline he insisted on a re-visioning of enormous proportion.

> *"Every few hundred years in western history, there occurs a sharp transformation….Within a few short decades society rearranges itself – its world view; its basic values; its social and political structure; its arts; its key institutions. Fifty years later there is a new world."*

In April of 1993, the same month that Hasan Ozbekhan and Alexander Christakis sat down to "search for the core of the Problematique," Peter Drucker delivered the keynote address at the SEI Center for Advanced Studies in Management at the Wharton school. Ozbekhan who was on the faculty at Wharton and Christakis attended. Drucker suggested in that lecture that *"In 2010 and 2020 it might be very hard for anyone to imagine what the world was like in 1970 or 1980. It is changing so fast."*

Christakis recalls Drucker acknowledging that Hasan Ozbekhan had been among the most prescient anticipators of the change that he too saw forthcoming. It may well be that Drucker saw Ozbekhan as the "educated person" that he considered essential to the New Realities.

The Scientific revolution had given us the expanded view of the world. However, its predisposition to mechanics and the predictability of mechanics coupled with the rise of industrialization had created in the modern Western consciousness assurances that Heidegger characterized as *"Cartesian thought annexed to the will to power"* (Barrett).

This sense of assurance manifests itself in cultural assumptions that the world, including the social world, was a system of relatively few unchanging but constantly improving parts whose relationships could be re-engineered by experts to refine a situation toward an ever improving equilibrium. This produced a confidence that history was inherently progressive and that if we became more informed about its trends we could continue to adjust and improve its equilibrium.

This enlightenment-based confidence in mechanics and its social manifestations had produced a culture of expertise which has proven over time to have been a form of narcissism that is pragmatically unsustainable. This brief epilogue is not the place to properly explore the convergence of American Pragmatism through the work of West Churchman as reflected in "The design of Inquiring Systems." However, we would be remiss if we did not address the profound influence of West Chuchman on Hasan Ozbekhan, and I suspect also on Peter Drucker's notion of the educated Person.

I want to bring Drucker's insight into the educated person into this reflection and then to contrast it to the somewhat beleaguered tone of Meadows and Randers in the light of Drucker's observation. I want to connect Drucker once again to the imagination and determination of Ozbekhan and Christakis.

Here is Peter Drucker:

> *"Post Capitalist Society deals with the environment in which human beings live and work and learn. It does not deal with the person. But in the knowledge society, into which we are moving, individuals are central. Knowledge is not impersonal like money. Knowledge does not reside in a book, a data bank, a software program: they contain only information. Knowledge is embodied in a person; carried by a person; created augmented or improved by a person; applied by a person; taught and passed on by a person; used or misused by a person. The shift to the knowledge society therefore puts the person in the center. In so doing, it raises new challenges, new issues, new and quite unprecedented questions about the knowledge society's representative, the educated person.*
>
> *The educated person we need will have to be able to appreciate other cultures and traditions… Tomorrow's educated person will have to be prepared for life in a global world… But he or she will have to draw nourishment from local roots and, in turn, enrich*

*and nourish their own local culture… The educated person will therefore have to learn to live simultaneously in two cultures, that of the intellectual who focuses on words and ideas and that of the manager who focuses on people and work…*

*We neither need nor will we get "Polymaths" who are at home in many knowledges; in fact we will probably become even more specialized. But what we do need – and what will define the educated person in the knowledge society – is the ability to understand the various knowledges."*

And Hasan Ozbekhan

*"The future is profoundly different. Here the mind does not encounter given happenings to limit and guide it. It must, so to speak, fill the whole vast and empty canvas with imaginings, with wishes and goals and novel alternative configurations that somehow possess reality and represent shared, or at least shareable values…*

(Hasan Ozbekhan; "Futures Creation")

And Alexander Christakis

*"Trying to invent the mathematics of Ekistics, coupled to the extravagant seminars of distinguished scholars, led me to the belief that there was something fundamentally wrong with the dominant social systems designing paradigm. It became clear that unless the social systems designing paradigm espoused the democratic ideal of stakeholder participation in the designing process, it could not become effective in reversing the dismal trends visible today in so many social systems. The dominant paradigm was as ineffectual in terms of resolving the acute problems presented in the Problematique, as was Ptolemy's paradigm in explaining and predicting the motion of planetary bodies around the sixteenth century."*

(Alexander Christakis:" A Retrospective Structural Inquiry of the Predicament of Mankind")

---

## On Deliberative Democracy

The practice of science is essentially the establishment of correlations that are tested and subject to the wisdom of the community of similar interests and skills. Occasionally there is a quantum shift such as the work of Copernicus and Darwin that is so compelling that it is both transformational and rapidly assimilated.

The practice of democracy need not be dissimilar. Now, it too can assemble correlations and test them for their efficacy with the wisdom of the community. Systems Design Science takes us in this direction but Structured Dialogic Design gives us an instrument with which to refine and enhance that work.

This suggests that for the first time surprises may be moderated. Planning in the tradition of West Churchman and Hasan Ozbekhan may be enhanced.

It may be that Structured Dialogic Design and similar methodologies will demonstrate capacity to invoke and provoke insights in the social sphere that may be of a Copernican nature and that in all likelihood would have been missed without it. Christakis intuitively understood this when he recognized that the discovery of the notion of Problematique and the subsequent discovery that the deep driver for its resolution was values-based is of historic significance.

In the final analysis the Club of Rome may have given us four gifts of which its current members are in all likelihood unaware:

The first is the profound expression of Meadows and Randers of the devastating challenge that limits represent to those enthralled by technique.

The second and third are the observations of Alexander Christakis and the core insights from "The Predicament of Mankind" that a Problematique cannot be reduced, and that the deep driver for its resolution is values-based. This is no small insight when you consider it in the context of Max Weber's insights into the consequences of social constructions that are not values-based.

The fourth contribution is that of Structured Dialogic Design which has given us the capacity to practice deliberative democracy and thereby bring to the social sphere a discipline closer to the rigors of scientific inquiry than was heretofore imaginable.

Structured Dialogic design has given us an instrument through which we can begin to hear from and even invoke from the wisdom of groups the conscious, unconscious, and not-yet-conscious insights from our cognitive, evaluative, and cathartic ways of knowing

## The Gift

The story is told of Alexander Christakis who was asked, "*If you could ask Socrates one question, what would it be?*" Christakis replied, "*I would ask him how he came to know that he did not know?*"

Structured Dialogic Design is about engaging the world and its future in the order of questions and not in the order of answers. It is about inquiry, not declarations. This is not to say that it is not action based. It is. But it is action based after doing the hard work of establishing and recording high degrees of correlation among a broad range of participants in open deliberative processes that are documented and may be iteratively revisited.

Implicit in the Axial age is a caution about arrogance and a respect for relationships and intuitive ways of knowing. It suggests the axial shift, implicit in the "Limits to Growth" and the "*Growing irrelevance of traditional values and continuing failure to evolve new value systems*" that Christakis and Ozbekhan found to be the core of the Problematique, is an inquisitive and authentic hope; an as yet little known gift, if you will, from the founders of the Club of Rome to posterity and its planetary home.

## Postscript

"*I walked into the Agora and was struck by the overwhelming feeling that agoraphobia was not just the fear of open spaces. It was the fear of openness.*"     Avery Manchester

Craig Lindell

## General References

Christakis, A. N. (1988). The Club of Rome revisited in: *General Systems.* W. J. Reckmeyer (ed.), International Society for the Systems Sciences, **Vol. XXXI**, pp. 35-38, New York.

Christakis, A.N. (2006). A Retrospective Structural Inquiry of the Predicament of Humankind Prospectus of the Club of Rome. Chapter 7 in Rescuing the Enlightenment from Itself: Critical and Systemic Implications for Democracy (Janet McIntyre, editor. Springer Science Business Media, Inc.

Christakis, A.N. and Bausch, K.B. (2006). *Harnessing Collective Wisdom and Power to Construct the Future.* Greenwich, Information Age Publlishing

Churchman, C.W. (1979). *The Systems Approach.* New York: Delta

Doxiadis. C. A., (1968). *Ekistics: An Introduction to the Science of Human Settlements,* Hutchinson of London.

Flanagan, T. and Christakis, A.N. (2010.) The Talking Point: Creating an Environment for Exploring Complex Meaning. Charlotte NC, Information Age Publishing

Judge, A. (2010). Club of Rome Reports and Bifurcations: A 40-year Overview. www.laetuspraesens.org/links/clubofrome.php . Retrieved 5/24/2020.

Masini, E.B. *The Legacy of Aurelio Peccei.* Club of Rome, European Support Center

Meadows D. H., Meadows D., and Randers J. (1972). *The Limits to Growth.* New York: Universe Books.

Özbekhan, H. (1969). Towards a general theory of planning. In E. Jantsch (ed.), Perspectives of planning. Paris: OECD Publications.

Peccei A. (1969 ). *The Chasm Ahead,* Toronto: The Macmillan Company.

Radzicki , M.J. and Taylor, R.A. U.S. Department of Energy. (1997). Origin of System Dynamics: Jay W. Forrester and the History of System Dynamics. Retrieved    5/27/2010.

Whitehead, J.R. (date unknown) *A Brief History of the Club of Rome: A summary and Personal Reminiscences. http://www3.sympatico.ca/drrennie/CACORhis.html*

Wikipedia. *Aurelio Peccei.wikipedia.org/wiki/Aurelia_Peccei.  wikipedia.org/wiki/ AureliaPeccei/ Retrived 5/27/2010.*

Wikipedia. Jay Wright Forrester. http://en.wikipedia.org/wiki/Jay_Wright_Forrester.  Retrieved 5/27/2010.

**Appendix 1.  Guidance on Learning Platforms for Instructors**

The individual components of our ad hoc learning platform are considered below:

Email

Course participants were engaged using individual email accounts, some of which may be supported through their home universities. Email was used to guide students to registration processes, to present class schedules, and to distribute instructions for accessing other components of the learning platform. None of our course participants reported difficulty with their email communications; however, high volume use of email alone as a means of exchanging and contributing to rapidly updated information is impractical in even modest size classes.

Voice-over-Internet

SKYPE software provides free, voice-over-Internet communication (http://www.skype.com/). Users need to have personal computers that include microphones and speakers and need to have administrative control over the computers that they are using so that they can download and install free SKYPE software. We have used this resource for groups of up to 16 participants. At the start of the course, email instructions for acquiring a SKYPE account were presented to students, and faculty SKYPE account names were shared. Students and faculty established individual calls amongst themselves in anticipation of an initial conference call. The initial conference call convened the class to elicit collective reflection of the course design, review of the syllabus, and questions related to assignments. This call also allowed class participants to discuss other components of the online learning platform.

SKYPE additionally provided students with an instant means of seeing when their instructor might be available for an impromptu "office visit" or when fellow students might be available for an impromptu consult. Easy voice contact adds an important mechanism for working with students who may be participating from different cultural and linguistic backgrounds.

Wiki Website

A jointly authored website was used as a repository of course participant contributions to the content of the course. Wikispaces was selected as the online repository for the course because it was judged to offer a facile system for managing multiple streams of threaded discussion, it had proven to be reliable in prior testing, and it offered its services without user fees. A class worksite was established and was sequentially expanded as the course progressed through its six week cycle (http://predicament-retrospective.wikispaces.com/).

The class wiki workspace was configured to provide distinct "workspace sections" for each of the following phases of class activity (though not all phases were used in this pilot course):
- the problem sets and their clarifications
- the students' individual preferences for most important problems
- the class's collective understanding of interactions among highly preferred problems
- the students' individual narrative accounts of that understanding
- the students' individual recommendations for acting on highly preferred problems
- the overlay of actions on the class's understanding of the system of influence among problems
- individual student reflections on the content and process of learning through this experimental online course

The class wiki workspace content included supportive documents in the form of:
- a disclaimer which clarifies that the class wiki is not a work product of the current Club of Rome
- a catalogue of key email notices about administrative issues within the course
- a record of the course announcement
- a library of course readings
- a syllabus of course tasks
- a list of course participants and their contact information
- a page providing world time zones to support in trans-global synchronous meetings

The Wiki workspace approach was based upon practices developed and validated by Gayle Underwood, who has 15 years of experience in education in online learning projects. She is the senior technology integration consultant for the Allegan Area Education Service Agency and is recognized for her leadership in Universal Design for Learning in Michigan schools. Internationally, Gayle has been supporting online learning for Turkish and Greek communities in the island of Cyprus and is working with Americans for Indian Opportunity (AIO) and the Advancement of Maori Opportunity (AMO) to enhance interaction and communication among indigenous people throughout the world. The effective use of this wiki, including orientation and coaching for course participants, is a task of the instructional staff.

Online Screen Sharing

Student access to online screen sharing involves responding to an invitation to enter a specialized, interactive website. For the purpose of this pilot course, a no-cost, trial membership was secured from GoToMeeting (http://www.gotomeeting.com/fec/). Students were emailed a URL for the website with instructions for entering the classroom and a time for signing into that website. The classroom can be open for public participation or password protected for private meetings at the instructor's option.

From the instructor side, software needs to be downloaded and a hosting session needs to be scheduled and launched. The online class used only basic features of the virtual classroom to enable online screen sharing. SKYPE conference calls were established concurrently to assure than any potential loss in connectivity with the virtual classroom could be recognized immediately. Serendipitously, it was discovered that agreement and disagreement with the significance of proposed influence among pairs of problems could be recorded in a facile fashion using the "chat" subroutine of the SKYPE product. This proved to be superior to a role call for oral votes for each relational assessment.

The virtual classroom proved effective as a means of sharing a software display screen as students engaged in real-time, pair-wise comparison of continuous critical problems.

Systems Structuring Software

Instructors applied CogniScopeII software to collect and display the 49 continuous critical problems, construct affinity clusters, and construct an interpretive structural model (ISM) based on the class's pair-wise comparisons. The class successfully constructed a tree-like map based upon their highly preferred CCPs. An academic version of this software package that is limited to mapping 15 problems is available free to academic users (see http://www.globalagoras.org/).

## Appendix 2.  Representative Four-Week Syllabus for Global Problematique Study

Upper Level / Graduate Level Syllabus for A Democratic Approach to Sustainable Futures

REGISTER FOR THE COURSE

Detailed instructions will be communicated through eMail unless otherwise stated by your instructor

        Your instructor will provide you with a website address (a URL) for the classroom wiki

        Your instructor will assign (or confirm your selection of) an online study team of 3 or 4 classmates

        Your instructor will provide email contact for study team members

        Your instructor will assign you your individual set of "Continuous Critical Problems" (CCPs)

        Your instructor will provide you with the date, time & website address for the online classroom

Your work will be individually turned in to the instructor unless specifically required as a wiki posting

        You will generate a research synopsis on your set of "Continuous Critical Problems" (CCPs)

        You will keep a "journal" of your reflections as you experience this online study program

        You will generate a "narrative" of the consensus discovered by the class

FIRST WEEK __ Getting onto the online spaces and taking a brief historic reflection

Download SKYPE www.Skype.com and create a user account (your computer must have a microphone)

        SKYPE is a voice-over-internet system providing you with instant access to teachers and classmates

        Email the user name that you created for your free SKYPE account to your instructor

        Make test call to instructor (your instructor must provide the SKYPE call address to you to do this)

        Add your Instructor to your SKYPE "contacts" list (so that your instructor's calls will reach you)

        Add member(s) of your online study team to your SKYPE "contacts" list and call that class member

        Connect with SKYPE to all members of your online study team

        Set up a time each week when your study team can all connect in a SKYPE conference call

        Email your instructor if you are experiencing problems with this technology.

# A Democratic Approach to Sustainable Futures

Sign in to classroom wiki (for example see:  http://predicament-retrospective2011.wikispaces.com/)

> The wiki provides "blackboards/whiteboards" and filing space for the virtual classroom
>
> Explore the wiki: the "navigation" panel on the left & each section has a set of tabs for pages
>
> Find your "profile" on the wiki and post your SKYPE username, a brief bio, and a statement about your hopes/expectations for the class (add a photo, if you choose)
>
> Post a "Comment" on the Home page confirming that you have updated your profile
>
> Note: Comment boxes are like email notes.  They are "threaded" in a stream of related comments
>
> Enter your name as a "new post" on the "discussion" page of the "home" section of the wiki
>
> On the "new post" you have just created, comment on the experience of getting started with the wiki
>
> Email your instructor if you are experiencing problems with this technology.

Read CH 1 of <u>A Democratic Approach to Sustainable Futures</u>

> "Engaging Global Sustainability as the Predicament of Mankind"
>
> Set up a SKYPE conference call with your online study team and discuss the 5 "study questions"
>
> Create a journal record of your study team's conclusions
>
> Individually begin researching your set of assigned "Continuous Critical Problems" (CCPs)

Come to the Online Classroom (see www.GoToMeeting.com or another of the online classroom companies)

> Our first class will discuss our access to SKYPE and to the classroom wiki
>
> We will confirm the status of our readings and be certain we understand the research mission
>
> CH 2 is a very, very complex chapter, and can only give you hints about the design of the approach

**End of Week 1**

# A Democratic Approach to Sustainable Futures

SECOND WEEK __ Understand an approach to highly complex problems

Read CH 2 of <u>A Democratic Approach to Sustainable Futures</u>

"Applying Structural Inquiry to the Predicament of Mankind"

Create a journal record of your "clustering" of <u>Inadequate Shelter and Transportation ???</u>

With your study team test the discussion question about "… understanding how a pair of systems scientists used some specialized approaches …" and create a journal record of your reflections on this exercise

Create a journal record of your narrative for the "tree" created by Hasan and Aleco

Read CH 3 of <u>A Democratic Approach to Sustainable Futures</u>

"Understanding Structural Inquiry as a Design Process"

Set up a SKYPE conference call with your online study team and discuss "Voting is usually a tool for delegation and rarely a tool for direct action" [you do NOT have to agree with this view]

Create a journal record of your study team's conclusions

Skim CH 6 of <u>A Democratic Approach to Sustainable Futures</u>

"Investigating Continuous Critical Problems"

Conclude researching your set of assigned "Continuous Critical Problems" (CCPs)

From your research notes, extract the essence of your research in a single paragraph

Navigate to the "49 Critical Problems" Section of the wiki

On the page located under the Discussion tab of this section, locate the "+ New Post" button

Click "+ New Post" and add the number and the name of one of your CCPs in the subject line

Insert your one paragraph summary of that CCP along with 2 website (URLs) from your research

BE CERTAIN TO CHECK THE "MONITOR THIS TOPIC" BOX (so you will get the replies)

Post, and then repeat this step for your other CCPs.

Come to the Online Classroom (www.GoToMeeting.com or a similar online classroom company)

A Democratic Approach to Sustainable Futures

We will discuss our reflections on ideas of complexity. Is the world getting still more complex?

We will discuss posting our research, and our need to clarify questions that will arise.

We will discuss the idea of clarifying questions versus posting exhaustive/exhausting reports.

**End of Week 2**

THIRD WEEK __ Question, Clarify and Understand Each Other's CCPs

Work with the classroom wiki – be certain to "sign-in" so that you can post comments

Read all 49 posts for the full set of CCPs

Identify 5 posts that you feel need to be clarified so that you understand them

At the bottom of the post, present your question for clarification as a "Comment"

BE CERTAIN TO CHECK THE "MONITOR THIS TOPIC" BOX (so you will see the response)

Respond (with a "Comment") to questions posted for your set of CCPs

NOTE: If you get several related questions for a specific CCP, you need only post a single reply

With your SKYPE team weekly conference call, discuss experiences with the clarification process

Create a journal record of your study team's conclusions of the clarification process

If assigned by your teacher, Read CH 4 of <u>A Democratic Approach to Sustainable Futures</u>

"Using Structural Inquiry for Individual Systems Thinking"

Download CSII software from the Agoras website (www.GlobalAgoras.org)

Follow the step-wise instructions in CH 4 using the 5ccp.rcmd file

Create a journal record of your individual use of the software

Print out your map of influence among the 5 CCPs that you structured

Hold a SKYPE conference call with your online study team to compare experiences

Create a second journal record of your study team's conclusions

Come to the Online Classroom (see www.GoToMeeting.com for one of the online classroom companies)

Discuss the clarification process in general

Discuss specific CCPs when ANY individual feels more clarification is needed

NOTE:  If a student cannot provide more detail, this should simply be acknowledged as a limit

As appropriate, discuss the map of the 5 CCPS that were individually constructed.

NOTE:  Did all students happen to map the same "deep driver."  How did maps differ?

**End of Week 3**

FOURTH WEEK __

If not assigned previously, skim CH 4 of A Democratic Approach to Sustainable Futures
"Using Structural Inquiry for Individual Systems Thinking"

Create a journal record of your understanding of this chapter

Read CH 5 of A Democratic Approach to Sustainable Futures: "Using Structural Inquiry for Group Problem Solving"

If you are meeting in a face to face classroom, you can cluster the 59 CCPs together

If you are working in a virtual classroom, you will be completing partial clusters

In a SKYPE online team conference call, review and cluster the last 10 CCPs

Label the clusters

Individually nominate five CCPs that you feel are of the highest importance

Tally the votes and record the preference voting results

Hold a SKYPE conference call with your online study team to compare experiences

Create a journal record of your study team's conclusions

A Democratic Approach to Sustainable Futures

Come to the Online Classroom (see www.GoToMeeting.com for one of the online classroom companies)

Review and confirm final composition of clusters

Launch the CSII software with the 40ccp.rcmd file and structure highly preferred CCPs first

Access the wiki to see clarifications for CCPs (as needed during clustering)

Complete the structuring and then display and review the map

Compose a preliminary narrative in the group

**End of Week 4**

FINAL WORK PRODUCTS

Narratives and Reflections from the course

Prepare  and post an individual narrative of the group structure to the "Stories" section of the wiki

NOTE:  Identify and explain the implications of deep drivers in the map

Hold a SKYPE conference call with your online study team to compare final experiences

Prepare and post a brief journal record of your study team's conclusions

Submit an expanded final individual journal record of the course to your teacher

**Appendix 3: : Evaluating this Course**

This appendix provides the Global Sounding score sheets along with the Global Sounding Moral Code of **David Loye**. The thinking behind including this in this text book is that all classes should be assessed in terms of the impact of their content. This appendix provides a self-guided assessment of the course. It aims to measure how the course has led to new ways of seeing the world and new capacity for living in the world. Please offer feedback to your teachers and also feel free to email your findings to the Institute for 21st Century Agoras at INFO@GlobalAgoras.org or directly to David Loye at loye@benjaminfranklinpress.com.

## *The Global Sounding and How to Use It*

Nuclear overkill. Global warming. Environmental devastation—increasingly obvious is the question to what extent do the policies and actions of governments, corporations, religions, political parties, and all other organizations drive us ahead, check us in place, or drive us backward in evolution.

In the case of all that threatens our health and wealth as individuals, medical science has developed such useful tools as the thermometer for reading temperature and the sphygmomanometer for reading blood pressure. This makes it possible for doctors to diagnose what ails us and help heal us.

Can an equivalent tool be invented to diagnose and heal all that's becoming a life-threatening illness for our species? That is, as with use of the thermometer with individuals, can something comparable be developed to diagnose and help heal regressive groups, communities, regions, and nations?

To this end, behind the development of the Global Sounding, lies thirty years of work by the new field of evolutionary systems science toward development of a fully human, action-oriented theory of evolution, a half century of measurement science, and the proliferation of new measures of global health and well-being described in *Measuring Evolution: A Leadership Guide to the Health and Wealth of Nation* (Benjamin Franklin Press, 2007).

The name Global Sounding comes from Darwin's famous voyage of the Beagle, originally commissioned to circle the world to obtain *soundings* indicating peaceful harbors and safe channels..

Designed to provide the first simple, "user-friendly" measure of evolution, along the first left hand column, under **Levels of Evolution,** are fifteen basic levels and activities for measuring the impact of policies and action on human and planetary evolution.

The second column, under **Indicators of progression,** shows what the progressive science of many fields indicates are descriptors of positive (+) impact on evolution.

The third column, under **Indicators of regression,** shows what the progressive science of many fields indicated are descriptors of negative (−) impact on evolution.

The last column, under **Regressive policies,** level by level provides examples from recent and current experience in America and globally of the kind of policies that inexorably drive both ourselves and our planet toward sickness and destruction.

**For more information, see www.davidloye.com  for Global Sounding and Global Sounding Moral Code in Menu.**  Both measure and moral code can be downloaded for free reproduction and sharing from my website: www.davidloye.com. Scientific background, development, and applications are explained in Measuring Evolution: A Leadership Guide to the Health and Wealth of Nations and Bankrolling Evolution: A Program for a President (both Benjamin Franklin Press, 2008).

*Please send stories of your pioneering use of this new measure to davidloye@gmail.com.*

## *The Global Sounding*

| Levels of evolution | Indicators of progression | Indicators of regression | Regressive policies |
| --- | --- | --- | --- |
| Cosmic | Sustainability of complex life forms | Environmental devastation | Opposition to environmental action and global concern |
| Chemical | Gaia Hypothesis/ symbiosis | Environmental devastation | Opposition to environmental action and global concern |
| Biological | Health and Longevity | Environmental devastation | Opposition to environmental action and global concern |
| **BRAIN** | **Parental love and nutrition** | **Lack of love and nutrition** | **Minimize governmental support** |
| Psychological | Self-actualizing | Lack of fulfillment | Maximize defense, minimize growth |
| [Cultural] | High priority for arts | Low priority for arts | Minimize support for arts |

# A Democratic Approach to Sustainable Futures

| Social | Freedom and equality | Control and inequality | Maximize control and inequality |
|---|---|---|---|
| **Levels of evolution** | **Indicators of progression** | **Indicators of regression** | **Regressive policies** |
| **Political** | Democracy | Authoritarianism | Oligarchic probe toward authoritarianism |
| **Educational** | Capacity for learning and independent thinking | Curtailing of facilities for learning and independent thinking | Radical de-escalation for progressive education |
| **Technological** | Emphasis on technologies of actualization | Emphasis on technologies of destruction | Radical escalation for the military |
| **Moral** | Living by the Golden Rule | Power of greed and corruption | Living by the Brass Rule |
| **Spiritual** | Sense of identity with humanity and greater being | Slavery to materiality | Celebration of absolute power of wealth |
| **Consciousness** | Cognitive, affective, and conative scope | Curtailing of scope of mind | Devaluing scope of mind |
| **ACTION** | Encouragement of progressive social action | Repression of progressive social action | Encouragement of *regressive* social action |

From *Bankrolling Evolution: A Program for a President* and *Measuring Evolution: A Leadership Guide to the Health and Wealth of Nations*, by David Loye. Benjamin Franklin Press, 2007. © David Loye 2007

A Democratic Approach to Sustainable Futures

## *The Global Sounding User Form*

### *How to use it in business, government, politics, science, education, by writers, the media, nonprofit and religious organizations, and philanthropists and foundations*

The Global Sounding is designed to be used as a guide to the investment of time and money by decision-makers in all of the above and other situations calling for the *right* decision in our lives today.

Focus on whatever cause, project, or policies you are considering putting time and/or money into.

Then follow these steps using the Global Sounding User Form.

1. Write down cause, project, or policy in the bracketed blank at the top.

2. With this subject in mind, scan down through the Indicators of Progression (+) and the Indicators of Regression (-).

3. For each level or activity of evolution (first column), ask yourself whether by investing time or money into the proposal for testing (your entry in the blank at top) you are likely to influence evolutionary *progression* (second column).

Or evolutionary *regression* (third column).

Or neither.

Or the question at this level or activity for evolution is just not relevant in your case.

4. If it seems the proposal might influence evolutionary *progression*, in the fourth column enter a plus (+ 1) in the blank.

5. If it seems it might influence evolutionary *regression*, in the fourth column enter a minus ( − 1) in the blank.

6. If it is hard to guess whether it would be of influence either way, or otherwise seems irrelevant, enter zero (0) in the blank.

7.  At the end, add up your pluses, minuses, and zeros, and enter the result in the blank for the Global Sounding Total.

*From Measuring Evolution: A Leadership Guide to the Health and Wealth of Nations, by David Loye.  Benjamin Franklin Press, 2007.  © David Loye 2007. Please send stories of your pioneering use of this new measure to davidloye@gmail.com.*

*The Global Sounding*

*Blank Form for Universal Use*

[[_____]]

| Levels of evolution | Indicators of progression (+) | Indicators of regression (-) | Enter impact (+, -, 0) |
|---|---|---|---|
| **Cosmic** | Sustainability of complex life forms | Environmental devastation | _____ |
| **Chemical** | Gaia hypothesis/ symbiosis | Environmental devastation | _____ |
| **Biological** | Health and longevity | Environmental devastation | _____ |

145

# A Democratic Approach to Sustainable Futures

| **BRAIN** | **Parental love and nutrition** | **Lack of love and nutrition** | _____ |
|---|---|---|---|
| **Psychological** | Self-actualizing | Lack of fulfillment | _____ |
| **[Cultural]** | High priority for arts | Low priority for arts | _____ |
| **Social** | Freedom and equality | Control and inequality | _____ |
| **Economic** | Caring and fair | Uncaring and unfair | _____ |
| **Political** | Democracy | Authoritarianism | _____ |
| **Educational** | Capacity for learning and independent thinking | Curtailing of facilities for learning and independent thinking | _____ |
| **Technological** | Emphasis on technologies of actualization | Emphasis on technologies of destruction | _____ |
| **Moral** | Living by the Golden Rule | Power of greed and corruption | _____ |

| **Spiritual** | Sense of identity with humanity and greater being | Slavery to materiality | _____ |
| **Consciousness** | Cognitive, affective, and conative scope | Curtailing of scope of mind | _____ |
| **ACTION** | Encouragement of progressive social action | Repression of progressive social action | _____ |
| | ***GLOBAL SOUNDING TOTAL*** | | _____ |

*From Measuring Evolution: A Leadership Guide to the Health and Wealth of Nations, by David Loye. Benjamin Franklin Press, 2007. © David Loye 2007. Please send stories of your pioneering use of this new measure to davidloye@gmail.com.*

A Democratic Approach to Sustainable Futures

## *The Global Sounding Code*

**A translation of fifteen scientific and spiritual guidelines for evolution
into tenets for the long sought Global Ethic
for our species**

### *1*

Honor and respect rather than plunder and rape our
living environment.

[cosmic evolution]

### *2*

Honor and respect Earth as the Mother of us all

rather than sell her into slavery.

[chemical evolution]

### *3*

100 years of science tells us Earth can still provide the

resources for health for all of us if given a fair

and practical human distribution system.

Establishing health systems for the many rather than

the chosen few is the moral requirement,

and must be our goal.

[biological evolution]

**4**

100 years of science tells us that a healthy brain and

healthy mind depends on adequate food

and love in childhood.

Thus nurturing and expanding, rather than

starving and blighting, is the moral requirement

for the feeding of brain and mind,

which must be our goal.

[evolution of the brain]

**5**

Mind is the precious gift of life to be nurtured, enlightened,

and celebrated,  rather than orphaned at birth and

thereafter degraded and exploited.

[psychological evolution]

**6**

Culture is *the most precious gift of evolution,*

to be cherished and nurtured in the wonder of its diversity,

rather than seized, shucked and twisted to serve low ends.

[cultural evolution]

**7**

Over thousands of years freedom and equality have been

and will remain the most precious goals for evolution,

rather than just words to be used by power-mad

deceivers to sell control and inequality.

[social evolution]

**8**

The failure of three hundred years of capitalism and two hundred

years of communism starkly reveal the challenge.

Moving behond them, the economics of caring and fair

sharing with others we know in our hearts could

end the breeding of corruption, misery,

and war, is the moral requirement,

which must be our goal.

[economic evolution]

*9*

Democracy,

rather than the sham of the oligarchy of wealth,

the creep by stealth toward fascism,

or the prison of any other form of tyranny,

is a moral requirement, and must be our goal.

[political evolution]

*10*

Education for all,

rather than the fact of good schools for some

and bad schools or no schools for all others,

is a moral requirement, which must be our goal.

[educational evolution]

*11*

Develop technologies to strengthen and expand the voices

of caring and reason and the abundance of life,

rather than technologies that ever more

powerfully threaten to, and indeed then

are driven to, launch and kill life.

[technological evolution]

151

**12**

Differently stated but the same for all, the message

of thousands of studies of progressive science is exactly

the same as for thousands of years for progressive religion:

Do unto others as ye would have them do unto you—rather than

do it unto others before they can do it unto you.

[moral evolution]

**13**

Affirm our identity but also our responsibility in tandem

with a greater being, a greater reality, or a greater

mission, rather than fool yourself by claiming

some one and only holy being, idea, or cause

for your own low ends.

[spiritual evolution]

**14**

The most precious gift of evolution is consciousness.

Love, nurture, and expand it, rather than

automatically seek to narrow, blunt,

or otherwise slap it into shape

and seize the pitiful remnant

for your own low ends.

A Democratic Approach to Sustainable Futures

[evolution of consciousness]

**15**

For over 100,000 years the choice before our species

has been unchanging.  Either join those who take action

on behalf of the better world,  or fall back in shame or ignorance

with those willing to sell out to the highest bidder and

join the march to extinction.

[evolution of action]

From *Bankrolling Evolution: A Program for a President,* and *Measuring Evolution:*

*A Leadership Guide to the Health and Wealth of Nations*, by David Loye.

Benjamin Franklin Press, 2007.

**Appendix 4. Opportunities to Gather Online Learning Research from the Global Problematique Course**

The Democracy and (Global) Sustainability course provides elements of three classes of online learning in a combination of asynchronous and synchronous activities. First, the course provides didactic or expository learning as the instructor presents an account of prior efforts to address the challenge of global sustainability, instructions for creating a systems view thorough individual and collaborative design, and introduction to a catalogue of continuous critical problems. The class website provides an asynchronous online means of delivering text documents and accumulating new content for review by each individual student. Second, the course provides active, asynchronous learning where each student individually conducts library and/or online research to find current examples of continuous critical problems for global sustainability, presents their findings to the class by posting their work onto the online workspace, and then also by individually creating and posting a narrative of the relationship they see among the high priority problems in the problematique for global sustainability. Third, the course provides asynchronous online interactive learning where students challenge each others' posted statements of the meaning of continuous critical problems in clarification activities, and synchronous online interactive learning where students convene in a virtual classroom to share understandings of – and construct consensus assessments for – relationships between pairs of continuous critical problems.

**Course Delivery Options**

For the purposes of assigning the specific impacts of online instruction as a substitute or as an enhancement for traditional courses, the Democracy and (Global) Sustainability course can be delivered in traditional synchronous, face-to-face classrooms which include asynchronous student research activities, as a blended classroom with some tasks supported through online tools, or as a fully online course with greater or lesser use of asynchronous individual activity.

**Learning Metrics**

1. Formative

Student learning will be assessed using formative and summary testing. The formative test will ask students at the start of the course to name factors that they feel are critically important to address in planning policies to assure global sustainability. They will be asked to identify which of these factors they feel are most important and to offer a brief statement for their assessment. Students will also be asked what they feel is the best way for a group to make a shared decision about actions which need to be taken to assure global sustainability. Finally, students will be asked if they feel that their views are likely to be shared by other course participants from a different part of the world, and to offer a view as to why this may or may not be the case. At the close of the course, students will be asked to state the set of critical problems which they researched, to define where their critical problems were placed in to the systems view of the problematique constructed by the class, and to recall what critical factors their research factor most directly influenced in the structure. Students will also be asked once again what they feel is the best way for a group to make a shared decision about actions which need to be taken to assure global sustainability. And also once again, students will be asked if they feel that their views are likely to be shared by other course participants from a different part of the world, and to offer a view as to why this may or may not be the case.

2. Summative

Performance will be assessed by variance between formative and summative test results on an individual basis, and accuracy of summative test results on an aggregate basis. Formative test results on postulated continuous critical problems that were anticipated by individual students will be scored by content analysis with the set of 49 continuous critical problems as reported in the 1970 prospectus for the Club of Rome.

**Non-learning Metrics**

One common conjecture is that learning a complex body of knowledge effectively requires a community of learners (Bransford, Brown and Cocking 1999; Riel and Polin 2004; Schwen and Hara 2004; Vrasidas and Glass 2004) and that online technologies can be used to expand and support such communities. An exit survey of course participants will include an assessment of their view of the quality of the "community" in the class, where their response will be bracketed by naming and scoring a class that they felt exhibited exceptionally high level of community and by naming and scoring a class that they felt exhibited an exceptionally low level of community. Students will be asked to rate their agreement with the following statement on a seven point scale: "With reference to other classes that I have taken, participation in this class increased my sense of being part of a global community."

Another conjecture is that asynchronous discourse is inherently self-reflective and therefore more conducive to deep learning than is synchronous discourse (Harlen and Doubler 2004; Hiltz and Goldman 2005; Jaffee et al. 2006). Students will be asked to rate their agreement with the following statement on a seven point scale: "With reference to other classes I have taken, my independent research on the continuous critical problems that were assigned to me significantly enriched my understanding of the challenge of global sustainability."

**Instruction Parameters**

Different classes for the Democracy and (Global) Sustainability course will include conditions which are unique to distinct settings and may also include practices that are unique to specific instructors or specific task modifications. For purposes of comparison across Democracy and (Global) Sustainability courses and with other online courses taught elsewhere, conditions and practices will be recorded based upon guidance provided by the US Department of Education's Office of Planning, Evaluation, and Policy Development report on the Evaluation of Evidence-Based Practices in Online Learning: A Meta-Analysis and Review of Online Learning Studies (2009). This guidance names 51 factors which in recent meta analysis of the impact of online learning have been identified as moderators of student learning. Conditions of the course will be catalogued by the instructor and corroborated by student review at the conclusion of the course.

Practice aspects of the instruction of the course will be assessed by student evaluations of teaching fluency and by instructor self-assessments. The student evaluations will additionally include a survey of the students' familiarity with the learning platform used in the course, and with their comfort using the platform for each of the learning tasks during the course.

**Assessment of Group Learning Metrics**

The Democracy and (Global) Sustainability course tracks idea generation, conversion, and decision making time based upon internal documentation of "interactive management" practices. Metrics will be collected for quantitative and qualitative assessment of content posted to the class learning platform by course participants and will monitor the pace of and incremental time for decisions to be specified by the class during online interactive synchronous learning. The structure of the dialogue for the class will be interpreted in the context advanced by Hron and colleagues (Hron, A., F. W. Hesse, U. Cress, and C. Giovis. 2000. Implicit and explicit dialogue structuring in virtual learning groups. British Journal of Educational Psychology 70 (1):53–64.).

**Replicability of Findings**

In the interests of promoting deeper research on conditions and practices that enhance online learning, all content used in the Democracy and (Global) Sustainability course, including the interactive learning software, will be made available to scholarly      researchers under a collaborative                              research                                agreement

## ABOUT THE AUTHORS

Thomas R. Flanagan, PhD, was raised in Massachusetts, attended universities in Massachusetts and in Connecticut. He began his professional career life science and biomedical technology and worked in and founded high performance companies in New England and in Europe both in technology and in management roles. Tom's appreciation for decision-making in diversified research and development teams let him to advanced studies at the MIT Sloan School of Management, and subsequently to an opportunity to work with Dr. Alexander Christakis and to support co-laboratories of democracy in international business, government agencies, and community missions. Tom has taught classes in biology, chemistry, engineering, and management for university students and, online classes for practitioner training in participatory democracy. His published books include The Talking Point: Creating an Environment for Exploring Complex Meaning, (with AN Christakis; Information Age Publishing, Inc., Charlotte, NC, 2010) and CogniSystem™I User's Manual: A Step-by-Step Guide for Collaborative Design (Institute for 21st Century Agoras, Riverdale, GA, 2006). Previous use of the approaches presented in this workbook appear in: T Flanagan, J McIntyre-Mills, T Made, K Mackenzie, C Morse, G Underwood and K Bausch, 2011. An Online Course in Sustainable Democracy: A Group Decision Making Process (in press).

Kenneth C. Bausch, PhD, grew up in Ohio and received his BA in Philosophy from Duns Scotus College followed by four years of intensive theological studies at St. Leonard's College. He began his professional life as a Catholic priest of the Franciscan Order and has been a pastor, a high school teacher, an inner-city organizer working with street gangs and community groups, a counselor, a social service administrator, a real estate agent, a homebuilder, a contractor, a university professor, a research director, and an organizational consultant. In the course of 40 years of inquiry into the human condition, he has delved intensively into philosophy, orthodox theology, Eastern religions, social and political ideology, psychology, sociology, and systems theory. Ken also holds an MA in Psychology at the State University of West Georgia and a Ph.D. in Psychology from Saybrook University. After a brief stint as Executive Director at the Ashley Montague Institute in San Francisco, Ken took up the leadership role of the Institute for 21st Century Agoras. Ken has taught psychology, sociology and systems science at Mercer University, Perimeter College, DeVry Institute, Capella University) and is currently teaching an online course through Flinders University in Australia. His published books include The Emerging Consensus in Social Systems Theory, (Kluwer Academic/Plenum Press; 2001), Body Wisdom: Interplay of Body and Ego (Ongoing Emergence Press, Riverdale, GA, 2010), and Harnessing Collective Wisdom and Power to Construct the Future (with AN Christakis; Information Age Publishing, Greenwich, CT, 2006).

## ABOUT HASAN OZBEKHAN

Hasan Özbekhan, PhD (1921-2007), Professor Emeritus of Management at the Wharton School of the University of Pennsylvania, was son of the Turkish Ambassador to Italy. He was educated at Lycée Chateaubriand in Rome, studied Law and Political and Administrative sciences at the Faculte de Droit and the Escole Libre des Sciences Politiques in Paris, and earned First Class Honors at the London School of Economics as a Leverhume Fellow. At the Wharton School, Özbekhan was Professor of Operations Research and Statistics, Professor and Chairman of the Social Systems Sciences Department, and Professor of Management. Among his celebrated contributions, he was a member of the founding team of the Club of Rome and authored the original prospectus for The Club of Rome "The Predicament of Mankind." He was a tireless advocate of inclusive, normative planning and a champion for transformational design. From his 1968 report "Toward a General Theory of Planning, Özbekhan states "*... any change that is not a fundamental change in values merely extends the present rather than creating the future.*"

## ABOUT THE AGORAS

The Institute for 21st Century Agoras (AGORAS), an international nonprofit 501(c)(3) [EIN 03-0466448] education entity incorporated in the State of California in June 2002, is a membership organization composed of university-affiliated, independent, and corporate social system dialogue managers who promote and enable participatory democracy through the practice of authentic, large group, collaborative design. By corporate charter, the AGORAS will:

• Promote the idea of human connectedness and interdependence (the "global village")
• Promote democratic processes for addressing the problems and opportunities associated with global economic and political integration ("globalization")
• Promote the establishment of co-laboratories of democracy (also known as 21st Century Agoras)

   The AGORAS maintains an archive of field applications of SDD under the oversight of a corporate research director. Archives will be accepted from any and all individuals who use structured dialogic design for democratic social system services, and collaborative research proposals will be seriously considered from all professional organizations addressing complex challenges that can advance resolution of global economic and political isolation.

|INSTITUTE for 21st|
|CENTURY AGORAS|

info@globalagoras.org
www.globalagoras.org

160

www.ingramcontent.com/pod-product-compliance
Lightning Source LLC
Chambersburg PA
CBHW051216200326
41519CB00025B/7135